DISCARD

DISCARD

The Tree of

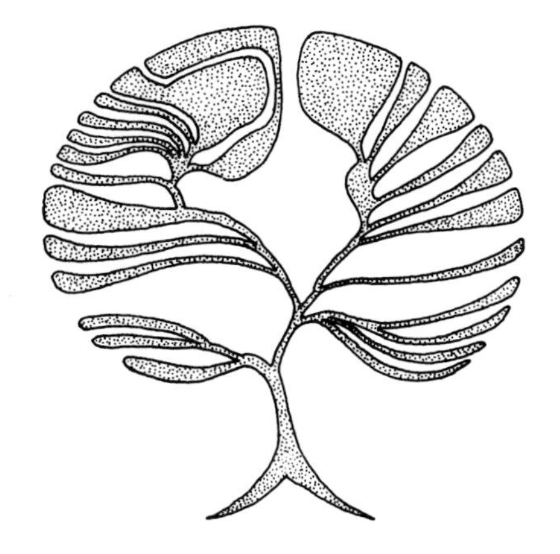

Animal Life

A TALE OF CHANGING FORMS AND FORTUNES

John C. McLoughlin

Dodd, Mead & Company New York

J
591.3
M

43947

THIS BOOK IS FOR ARIANA MERRILL McLOUGHLIN

ACKNOWLEDGMENTS

Special thanks are due Dr. Philip L. Shultz, surgeon, conservationist, and friend, who provided space, feedback, and encouragement throughout the preparation of this book; and my wife, Carole Crews McLoughlin, who keeps me in one piece.

Library of Congress Cataloging in Publication Data

McLoughlin, John C.
The tree of animal life.

Includes index.
Summary: Discusses the evolution of the major animal phyla from the first primitive cells and transformations of animal life by the processes of natural selection acting on genetic material.
1. Phylogeny. 2. Evolution. [1. Animals. 2. Evolution] I. Title.
QH367.5.M35 591.3'8 80-2789
ISBN 0-396-07939-3 AACR2

Copyright © 1981 by John C. McLoughlin
All rights reserved. No part of this book may be reproduced in any form without permission in writing from the publisher. Printed in the United States of America
1 2 3 4 5 6 7 8 9 10

Illustrations on the opposite page show Mesozoic and Cenozoic forms that inhabited similar econiches. Mesozoic animals are on the left, and the Cenozoic animals that replaced them millions of years later are on the right. (From top, left to right) *Aquatic air-breathers: mosasaur, whale; Flyers: early bird and pterosaur, modern bird and bat; Large herbivores:* Triceratops, *buffalo; Large carnivores: megalosaur, hyena; Small carnivores: coelurosaur, genet.*

The Tree of Animal Life

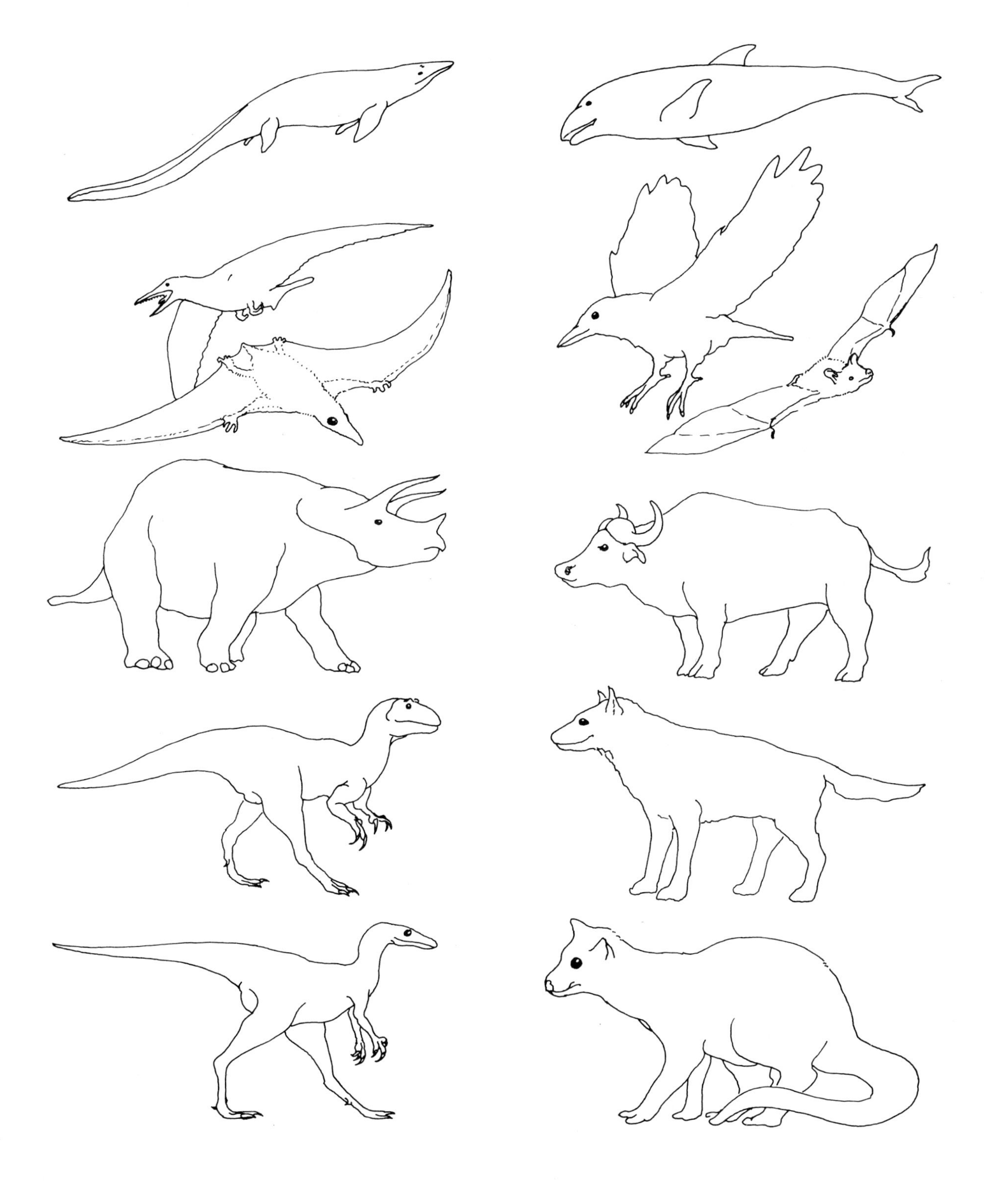

AUTHOR'S NOTE

People have always sought their origins, perhaps to gain insight into their futures. Until recently, the most ancient written record generally available to western Europeans was the Bible, which was carefully searched for matter pertaining to the origin of humanity. Using the Bible as his source, for example, an Anglican Primate of Ireland (Archbishop James Ussher) announced in 1654 that the entire universe was created at nine o'clock in the morning, Greenwich time, on 26 October, 4004 B.C., making the world 5,658 years old by his reckoning.

For about fifteen hundred years, the Bible has dominated European thought, and small wonder. Those who differed with its interpretation of the world were all too often publicly and most barbarically murdered as examples to other would-be disbelievers. With the rise of science during the past few centuries, however, we have gradually come to appreciate the empirical method of understanding, in which direct and repeatable observation, rather than "revealed" dogma, serves as the basis for much of our knowledge. The rise of science permitted such natural philosophers as the Englishmen Alfred Russell Wallace and Charles Robert Darwin to formulate the cornerstone theory on which is built the history of animal life as it appears in this book.

Here there are no ferocious gods, jealous and prejudicial, who create and destroy at whim. Instead, we find a stately tale of thousands of millions of years duration, a gradual unfolding of the universal principles that permit such unlikely beings as ourselves to exist. Based on observation of living forms and on analysis of traces (fossils) surviving from the remote past, we gain a rich and growing understanding of our origins and a look into the options for our destiny. The tool we use in interpreting the available evidence is called the Synthetic Theory of Evolution—"synthetic" because the original ideas of Darwin and Wallace are perpetually being augmented by more recent ideas and discoveries made available through the infant sciences of genetics, biochemistry, ecology, and paleobiology. The Synthetic Theory as a whole is an intellectual framework within which the observable conditions of the living world fit more consistently and precisely than in any other we have yet found.

A kangaroo sampler. (from left) *A potoroo, or rat kangaroo; a tree kangaroo; a wallaby; a great gray kangaroo.*

Consider the kangaroo.

For most of us, the great red kangaroos of the Australian grassland come to mind. These big jumpers, sometimes weighing a hundred kilograms, are the very symbol of Australia. Actually, more than fifty kinds of kangaroos inhabit Australia, ranging in size from the great red and his close cousin, the great gray kangaroo, all the way to the small rat kangaroo, which looks very much like a real rat and runs on four legs as well as jumps.

The familiar big kangaroos eat grass and other plants. In fact, they do it so well that in some parts of Australia they encroach on the grazing lands of domestic cattle and tend to be regarded as pests by the stockmen. Domestic cattle were introduced into Australia by European settlers within the past few centuries.

Leaves of grass are stiff and tough because they are composed largely of cellulose, the fiber that serves as the skeleton for the great majority of land plants. Most animals cannot digest grass, although it is very rich in food energy. And, because it is so common, it offers a great potential food source to any creature able to exploit it.

The stomachs of kangaroos and cattle digest grass by providing "vats" in which the protozoa that do the actual work may live. In these drawings, the vats are shaded; arrows show the direction of food through the stomachs.

Certain creatures do use cellulose as food. These are protozoans, or one-celled animals, as well as some bacteria. Mammals like kangaroos and cattle, on the other hand, cannot digest cellulose—not, that is, without help, which they get from the tiny protozoans and bacteria. These single-celled beings cannot live on the dry, sunbaked Australian savanna. But they can live in the water that cattle and kangaroos drink, and they enter those animals' bodies when cattle and kangaroos refresh themselves at their watering holes.

Inside those animals, the protozoans and bacteria find a moist, dark place where they can digest away at soaking wet, ground-up grass, for both kangaroos and cattle have evolved an ingenious solution to the problem of eating grass. Within their digestive tracts near their stomachs lie great fleshy vats, in which grass is stored after the kangaroos and cattle have moistened and chewed it with their grinding teeth. In these vats, the ground-up mass is held while millions of bacteria and protozoans live and work in comfort.

After the one-celled creatures have digested the grass to the right degree, muscles move it to the intestines of the cattle or kangaroos, where it is absorbed by the body. Thus, by the use of microscopic "middlemen," kangaroos and cattle are both able to eat grass very comfortably, and so it is that cattle growers in Australia are often not too fond of kangaroos.

Nevertheless, although kangaroos and cattle both live in the same sorts of places (open grasslands), eat the same food, use the same clever mode of digestion, and even travel about in large social groups that are similar to one another, they are not at all close relatives. In fact, cattle are more closely related to us and to whales and to bats than they are to kangaroos. How can two such distant relations as a cow and a kangaroo be so much alike in so many important ways? And what does it mean to say that animals as seemingly different as a whale, a bat, a cow, and a human being are close relatives? Herein lies a great tale, the tale of the growth of the tree of animal life. By considering the cow and the kangaroo, we will arrive at the beginning of that tale.

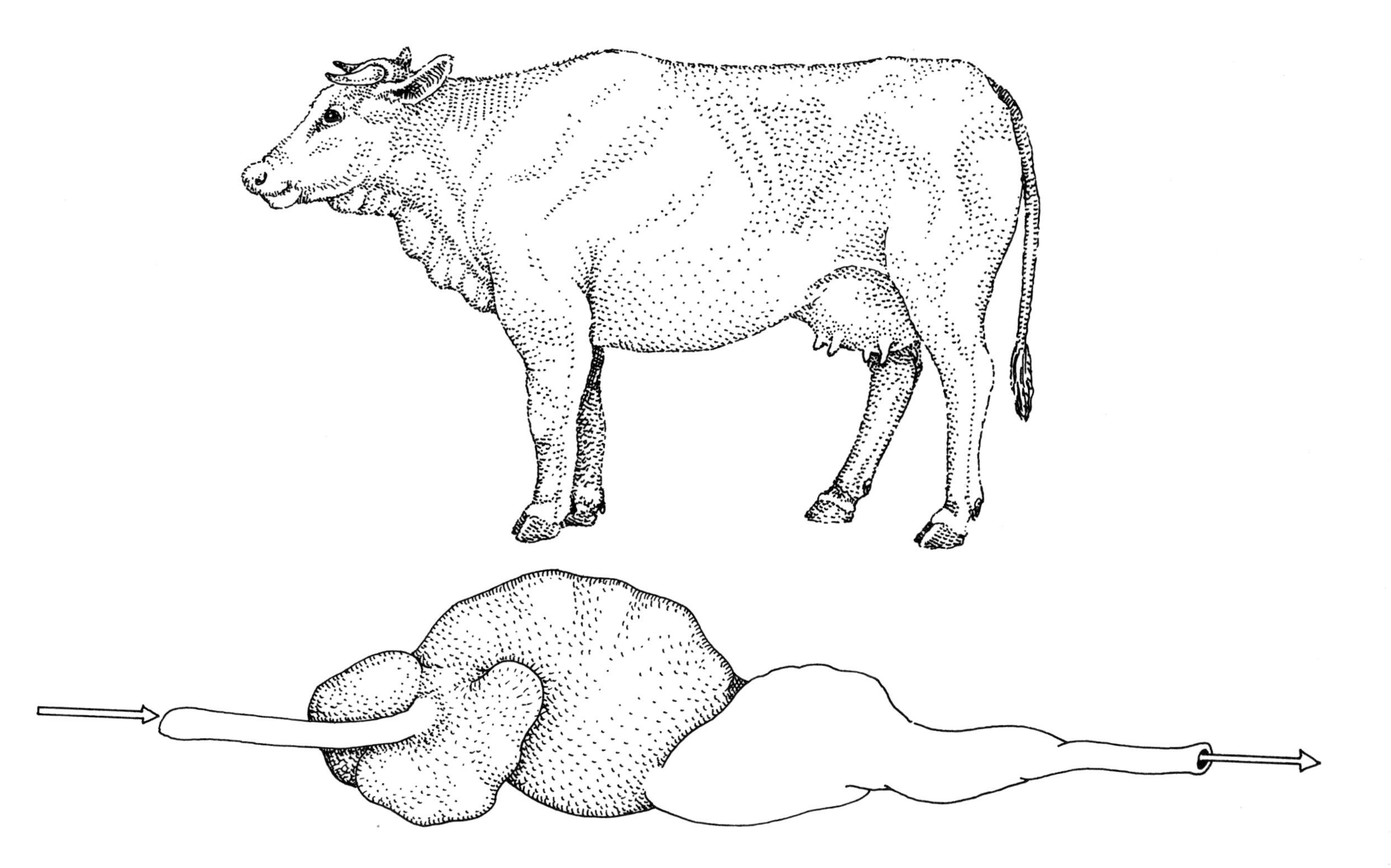

Both cattle and kangaroos have a common ancestor that lived about 100 million years ago. This shared relative was a mouse-sized furry sniffer called a pantodont. It ate insects and perhaps seeds—anything that would satisfy its gnawing hunger. It probably laid eggs and took care of them until they hatched. And, to give its offspring a better chance to survive in a world that was then dominated by all manner of hungry dinosaurs, the pantodont may even have cared for its young for a short while.

The pantodont family did moderately well for millions of years and then became extinct. Pantodonts were gone forever from the face of the earth. But they had left behind two new lines of little furry sniffers, much like the pantodonts themselves.

The two new sniffer families were different from one another mainly in the way they bore their young. Neither group laid eggs. In one group, the young developed inside their mothers' bodies until they were fully formed. Then they were born, as well developed as if they had just popped out of eggs.

The young in the other group developed inside their mothers' bodies until they could breathe air. Then, still immature, they were born and had to crawl to a pouch where they would attach themselves to their mothers' nipples and continue to develop.

The lowly pantodont and its descendants. (opposite page, above) *A primitive marsupial carrying her litter in a pouch, and a newborn, undeveloped at birth.* (below) *A primitive placental mammal nursing her young, and a newborn, fully developed at birth.*

The first of these sniffer families founded the line of placental mammals: animals like ourselves, cattle, whales, and bats, which nourish their young inside their bodies with special organs called placentae until those young are quite well developed. A calf, for instance, is able to walk at birth.

The other new clan, carrying their helpless embryonic young in pockets until they reached a point where they could move about and deal with the outside world, founded the line of marsupials, the pouch-bearers. Kangaroos, of course, are marsupials. Their young are born pink and hairless, sometimes no more than an inch long, barely able to crawl inside the mother's pouch to reach her teat.

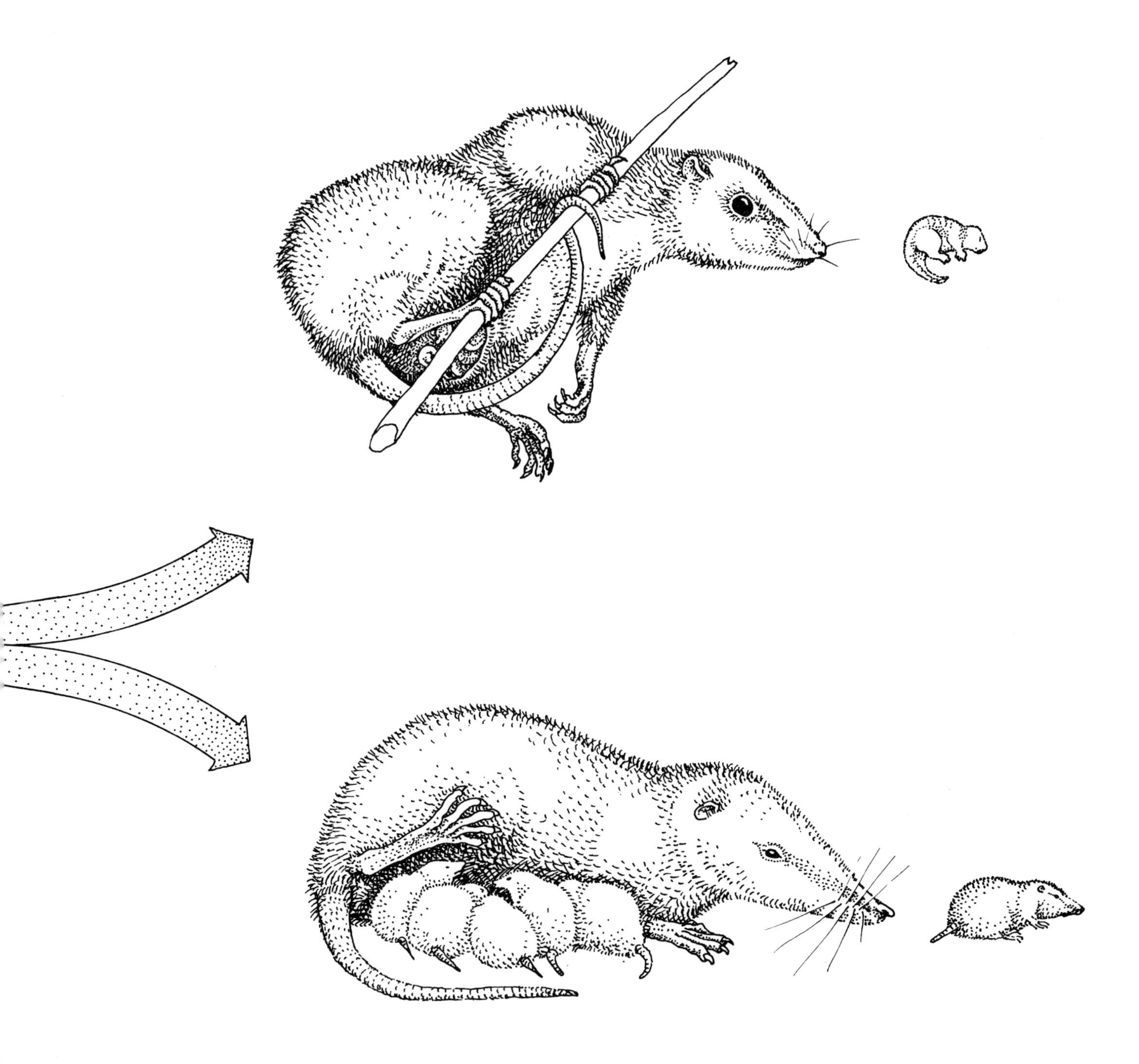

Both marsupials and placentals originated because laying eggs had caused trouble for the ancestral pantodonts. In that terrible world of 100 million years ago, pantodont eggs were often eaten while mothers were out trying to find food, and many a pantodont family came to a sorry end. This sort of relentless trouble, encountered by all living beings at some stage of their existence, is called selective pressure. It can take many forms. Overpopulation, which increases competition with others for limited resources, or climactic changes for which animals are not well adapted, are two common examples of selective pressure. Some form of selective pressure is always with us; we will run into it often in our tale.

The selective pressure operating on pantodonts acted through their eggs—something was always eating or stepping on or otherwise messing the eggs up. Once the young were hatched, they could follow their mothers about and learn how to get food for themselves. But oh! those eggs, those hungry dinosaurs! What to do?

Most young animals grow up to look and act very much like their parents. Now and then, however, one is born that differs from its parents in some distinct way. Perhaps its skin or fur has no pigment, or its hind legs are unusually large, or its beak is curved instead of pointed. When such a striking and unpredictable difference occurs among offspring, a difference that can be transmitted to the next generation, we say a mutant has appeared. It represents a spontaneous change in the genetic material of one of the parents that has been passed on to and expressed in the young. Such a change—a change that can be inherited—is called a mutation. Mutation may occur when a cosmic ray from space strikes the nucleus of a reproductive cell, or it may occur in a chance mix-up in the sequence of genes (hereditary information that dictates the character of a fully developed organism). Most mutations are either indifferent changes or are actually harmful to the organism. But a few—perhaps one in ten thousand—may produce changes that are useful to the animal in which they occur.

For the pantodonts, chance seems eventually to have provided two sets of useful mutations, both at about the same time and both allow-

Morganucodon, *one of the most primitive fossil mammals of which we are aware, lived about 185 million years ago. Shrewlike in appearance and probably in behavior,* Morganucodon *may have laid eggs; from a form very like this descended the live-bearing mammals, the marsupials and placentals.*

ing their descendants to evade once and for all the problem of fatally fragile eggs. The first marsupials and placentals looked and lived very much like their pantodont ancestors. But with the tremendous advantage of not having to leave eggs about, these two new clans quickly outnumbered their pantodont relatives. They pressed them into extinction and took their place in the world.

The "place" of the pantodonts, and of their placental and marsupial descendants, was not simply the physical space where they lived (on the forest floor and in the trees). Their "place" was really the whole way they fit into the community of plants and animals around them: the particular food they ate, when and how they got it, the size and kind of area they inhabited, the animals that hunted them, and other factors as well. Such a "fit" is called an ecological niche, or "econiche" for short.

The econiche of the pantodonts, and of the first placentals and marsupials, was one of eating insects and seeds in a dinosaur-filled night. Because the two new clans occupied the same niche as the pantodonts but were able to carry their immature young about with them, their young were more likely to survive. And so the placentals and marsupials flourished at the expense of their forebears.

Needless to say, all the little insect-eating marsupials and placentals were just as subject to mutations as their pantodont forebears had been, and, indeed, as were all living things. And such mutations did occur. As ever, perhaps one in ten thousand permitted any advantage to its possessor; but there was plenty of time and plenty of mutations.

While the dinosaurs were still common, our insect eaters were under pressure from these very efficient and well-designed creatures. Dinosaurs securely occupied all the econiches suitable for ground-dwelling animals larger than a modern pigeon. Their presence kept the early placental and marsupial mammals pigeon-sized or smaller (since anything larger was likely to get eaten), and so they remained during the entire rule of the dinosaurs.

Still, there was room throughout the world for little mammals, and they occupied their own econiches very well. By around 80 million years ago, the world's forests were probably teeming with mouse-sized mammals, both marsupial and placental. Because they occupied similar econiches, the two groups came into competition, producing for each other a new selective pressure similar to the one they'd exerted on their pantodont ancestors. Because the placentals bore their young at a more advanced, and therefore stronger, more agile stage of development, they had an advantage over the marsupials. Selective pressure began to work against the pouch-bearers.

For marsupials, those were hard times indeed. Again, though, chance was in their favor—this time not through mutation, but through a lucky accident in which perhaps one marsupial (a pregnant female) or two (a fertile male and fertile female) happened to be caught in a tree that fell into some water and drifted out to sea. For days they may have drifted, perhaps even for weeks, possibly still able to dig insect larvae or seeds out of their natural raft as they floated where the wind and waves would carry them.

At about this time, the earth's lands were joined into one large mass, with the exception of certain island continents, one of which would someday be called Australia. (It was still 80 million years before human beings would be around to name it.) Then, as now, Australia was separated from the rest of the landmass by a goodly stretch of ocean.

Ultimately, the marsupials' tree-raft brought them to this giant island. It happened that Australia in those days did not possess mammals of any sort (unless one would include the lowly monotremes, furry egg-layers more primitive even than the extinct pantodonts). There were dinosaurs in Australia, since it had been part of the world continent in the days when dinosaurs evolved. But mammals, which appear to have evolved in the northern part of the world continent, had not reached the southern areas by the time those lands began to drift off and become islands. The nighttime insect-hunting econiche was wide open in Australia.

Without other animals to compete with them in their econiche, the pioneering marsupials spread across their newfound continent in almost no time—"no time," that is, measured in terms of biological evolution, which is a very slow process. Fifteen million years later, the marsupials of the main continent to the north (a continent that would eventually break apart into Eurasia, Africa, and North America) were extinct, driven out by the more successful placentals. Australia's nighttime insect-eating econiches, however, were totally occupied by marsupials, descendants of the lucky one or two that had drifted there. (The comparable econiches of South America, also then an island continent, were accidentally populated by a very few marsupial drifters, too—but that is another story.)

Around 65 million years ago, something terrible happened to the entire earth, something that killed off nearly all the largest, most advanced animal life on the planet. This catastrophe, to which we shall return later in more detail, left vacant all the large animal econiches. The dinosaurs became extinct on land; in the oceans, the mighty marine reptiles died out. This period seems to have been one of starvation, in which big, active, hungry animals died, while small animals that could easily get about to new food sources (such as insects and birds) or that could go into temporary suspended animation (as can most reptiles and primitive mammals) survived.

Emerging from suspended animation, the world's little mammals did not encounter any large animals at all! There was no longer selective pressure from the dinosaurs, pressure that had kept the mammals small and insignificant. The rate of random mutation that affects all beings still operated on the mammals, only now those that happened to change toward, say, larger size, were *not* selected out of the breeding population by dinosaurs. Indeed, early large mammals were now pretty much kingpins in the world.

Mutations continued, and the larger mammals took to the econiches formerly occupied by the dinosaurs. Over millions of years, on the interconnected continents of Eurasia, Africa, and North America, dogs, cats, bears, and other meat-eating placental mammals filled the hunting econiches that had been left vacant by the carnivorous (meat-eating) dinosaurs. Multitudes of placental mammals that ate plants, such as elephants, horses, rhinoceroses, antelope, swine, deer, and cattle, evolved in the old econiches of the herbivorous (plant-eating) dinosaurs. These large mammalian plant eaters became food for the carnivorous mammals. In the seas, placental swimmers re-occupied the old reptilian econiches, becoming seals, walruses, whales, and manatees.

In isolated Australia, the identical drama was enacted, but with the empty dinosaur econiches being filled by the descendants of the marsupials that had colonized that huge island so long ago. The econiches were the same as on the main continent, and the animals filling them therefore shared many characteristics. Australia saw the rise of marsupial "cats" (dasyures), "dogs," and others that fed on the marsupial plant eaters that repopulated the landscape.

Some of these Australian predators became quite large and swift. One of the best known among them is the thylacine, or marsupial wolf, now very rare but once widespread across the Australian continent. The thylacine looks like a sort of wolf or a wild dog, and it pursues its game across the landscape in the same fast-running manner. The real (placental) wolf, however, is more directly related to us than to the thylacine, which so closely resembles it. Their resemblance is not one of similar family, but of similar econiche. The wolf and the thylacine are alike in form because the roles they play and the skills they must rely on parallel one another in their different environments.

Like the wolf, the thylacine is a good runner because generations of thylacines have survived by chasing swift prey. Like the wild cattle of the world, which must escape such predators as wolves, the kangaroo is speedy and rather belligerent if cornered because generations of kangaroos have survived by escaping such predators as the thylacine.

Over many generations, predators and their prey shape one another as they compete. The best jumpers among the kangaroos tended to escape the thylacines and live to reproduce. This eventually resulted in speedier kangaroo populations. Since only the swiftest, most powerful thylacines were able to catch the increasingly fast kangaroos, their populations also became faster and hardier over time, as the fittest among them survived. And if mutations happened to endow thylacines with traits that made them better runners, the kangaroos would be caught more easily—until some mutation occurred that produced a faster kangaroo. And so it went as our modern jumping kangaroos evolved, some of which can cover long distances at twenty-five miles per hour without tiring.

Periods of rapid change in which newly successful organisms begin to occupy all available econiches are called adaptive radiations. Having adapted through chance mutation to a new set of circumstances, a group spreads out (or "radiates," as do lines from the center of a circle). With time and successive mutations, the group is able to occupy an ever-increasing set of econiches—until some new selective pressure slows the process.

Wherever on earth they are, and in whatever evolutionary period they exist, similar econiches demand that certain characteristics of their occupants be similar. Whether or not these occupants are closely related, they will tend in some ways to remind us of one another. For instance, grass-eating mammals, wherever they exist, must es-

tablish a partnership with microscopic cellulose eaters. All known grazing mammals have evolved some sort of fermentation vats in which their food is broken down for them by those microscopic allies. The process by which unrelated animals that occupy similar econiches resemble one another is called convergence. We say their forms and functions are parallel.

Our kangaroo and cattle tale illustrates many of the primary principles of evolution that we will encounter in this book. Again and again, we will see an animal group overflow its econiche, experience some form of selective pressure, and, through some lucky mutations, come to circumvent that pressure and to experience an adaptive radiation into a new set of econiches, until competition within the new econiches becomes intense enough to begin the selective pressure over again. This cycle is central to the flow of evolution and to the growth of our tree. In fact, this same cycle continues today to dictate the patterns of human cultural evolution. Unless we take time to understand the evolutionary cycle, we human beings may very soon cease to exist, victims of our own unique selective pressures.

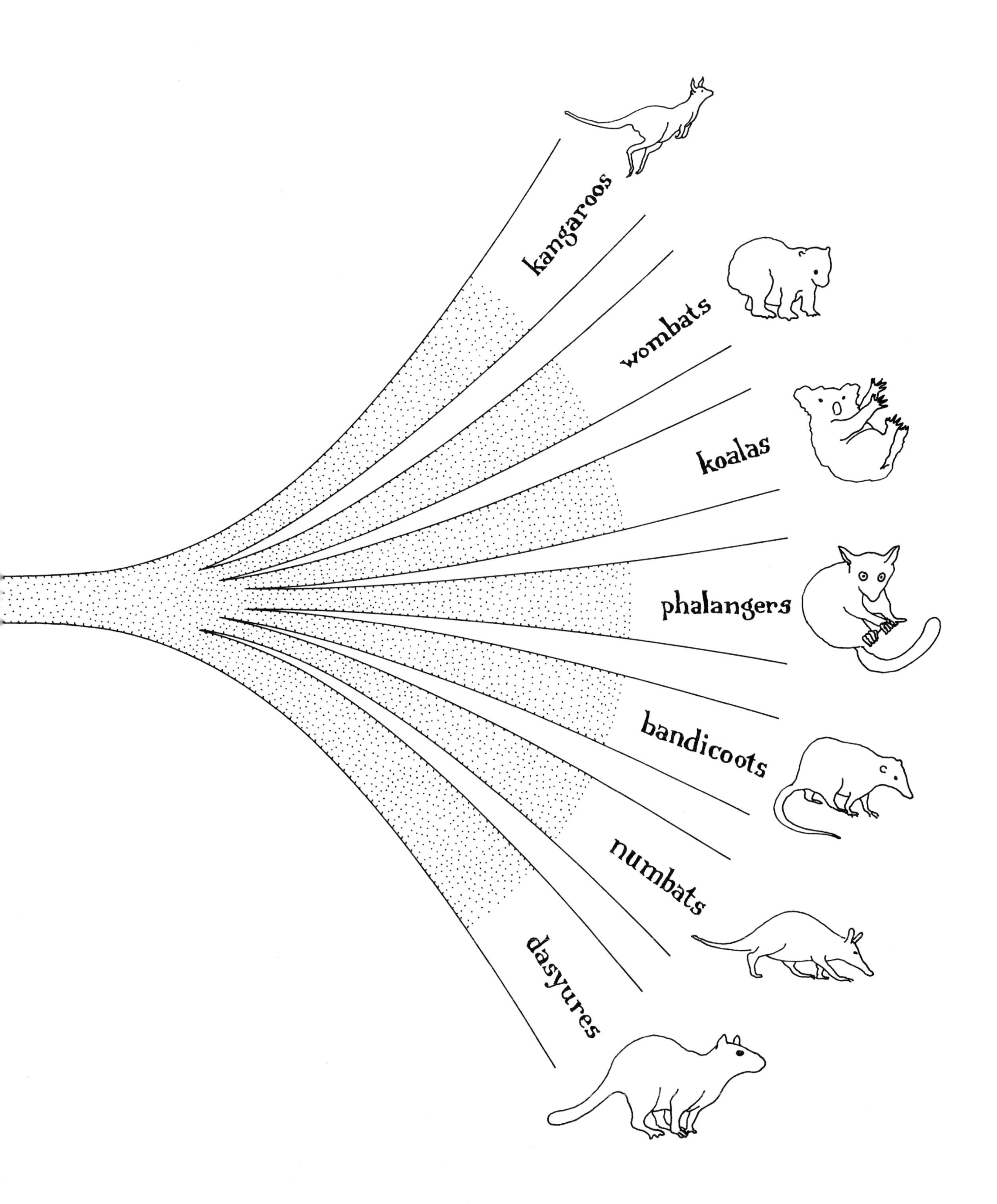

THE ADAPTIVE RADIATION OF THE AUSTRALIAN MARSUPIALS
FROM A COMMON PRIMITIVE ANCESTOR.

Life probably arose on Earth in a place like this, where minerals washed from the continents accumulated in a delta at the mouth of a river. Here, water, air, earth, and fire (in

The story of evolution on our planet is a long one, so long as to be virtually meaningless to most of us. The planet Earth is between 5 and 6 billion years old, a great ball of nickel-steel with a thin scum of lighter rocks, liquid and frozen water, and various gases covering its surface. In this respect, the earth is like Mars or Venus or Mercury: it is made of pretty much the same substances in about the same proportions.

Earth, however, is special in one very significant respect. It is just far enough from the sun so that at any given moment most of the water on its surface is liquid rather than frozen into ice (as on Mars) or evaporated into steam (as on Venus). By about 4 billion years ago, our planet was covered by oceans, much as it is now; about three-quarters of its rocky surface was submerged.

Liquid water is essential to the story of life on our planet, for it is a solvent in which most of the common substances of the earth's surface can be dissolved and mixed. The early earth was a turbulent

the form of lightning and sunlight) combined to produce the first living forms.

place, with storms lasting centuries, vast electrical discharges, and an endless flow of ultraviolet light from the sun. Water evaporated from the oceans into clouds, which were carried overland by sun-stirred winds. There they released their moisture onto the proto-continents in the form of rain, and the rain carried minerals back to the sea in an endless cycle that continues today, billions of years later. The planetary waters became a great stewpot in which all manner of elements were perpetually stirred by energy from the sun.

The immense chemical mix that resulted can be at least in part duplicated in the laboratory today. If we place some common substances such as water, methane, ammonia, and carbon dioxide in a vial and heat them, then flash sparks of electricity through them, and then cool them in a repeating cycle, the mixture quickly becomes tarry and black. On analysis, this mixture turns out to be composed of amino acids and other rather complex carbon-based organic molecules, the "building blocks" of life.

In the primordial oceans this happened on a much larger scale, to such an extent that the organic molecules dissolved in the water became ever more complex, dissolving and crystallising and dissolving again through the millenia. Crystallization is a process in which orderly structures grow as molecules add to themselves from materials dissolved in the water around them.

Most familiar crystals (the crystal of table salt is a good example) are simple molecules arranged in a structure that is rarely altered. One salt crystal is essentially identical to another. Among complicated organic molecules, however, different crystals might arise from mixtures of similar elements. The differences among them were reflected in the molecules' longevity. Those whose stability was greater tended to hold their form longer and to recrystallize more efficiently out of the primordial oceans.

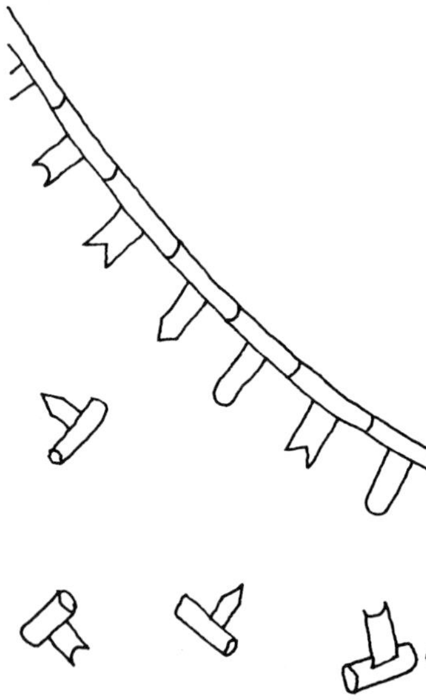

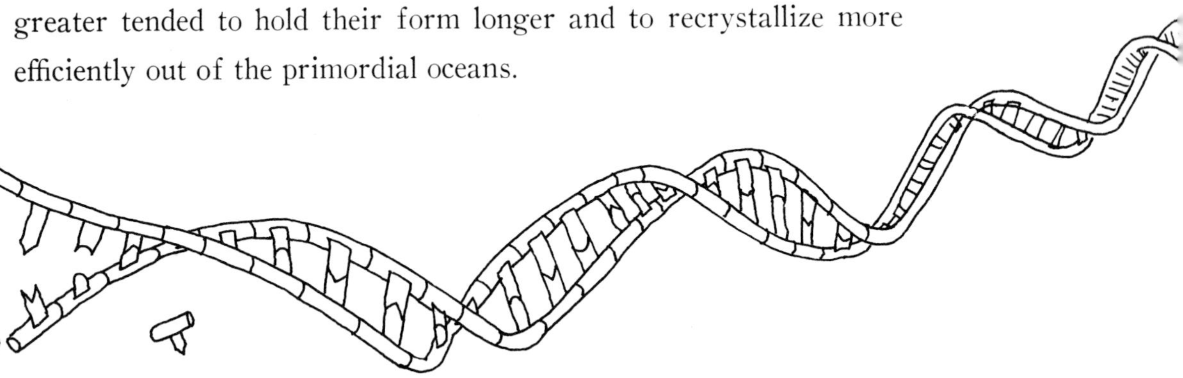

Somehow, complicated organic molecules came not only to be able to add to their structures, but to make complete copies of themselves—to replicate. And, because these molecules were large and complex, they sometimes made mistakes in replication. The molecules might not at all times make *identical* copies of themselves. This mistake-making became more pronounced as more complicated molecules came to exist in the soupy oceans of that long-ago time.

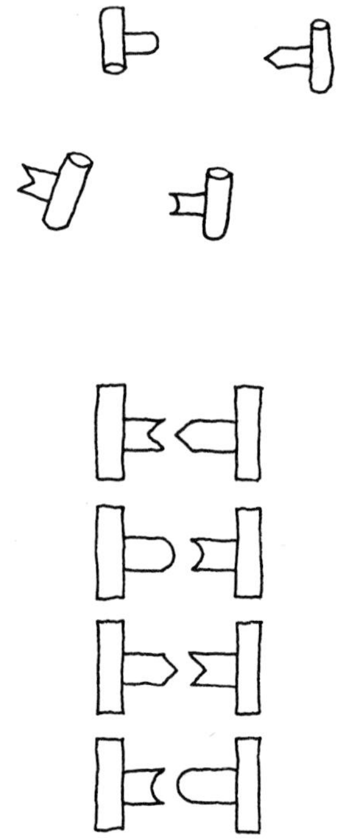

These early self-replicating, mistake-making molecules we know as nucleic acids, the most common of which is deoxyribonucleic acid, or DNA. DNA is the fundamental hereditary material for all living organisms. It carries the genetic information necessary to build a living being. This information, collectively referred to as genes, determines whether an organism is a man or a mushroom.

A single strand of DNA synthesizes its complement from a pool of free nucleotides, represented here by four types of little "keys" that make up the "alphabet" coding for living things.

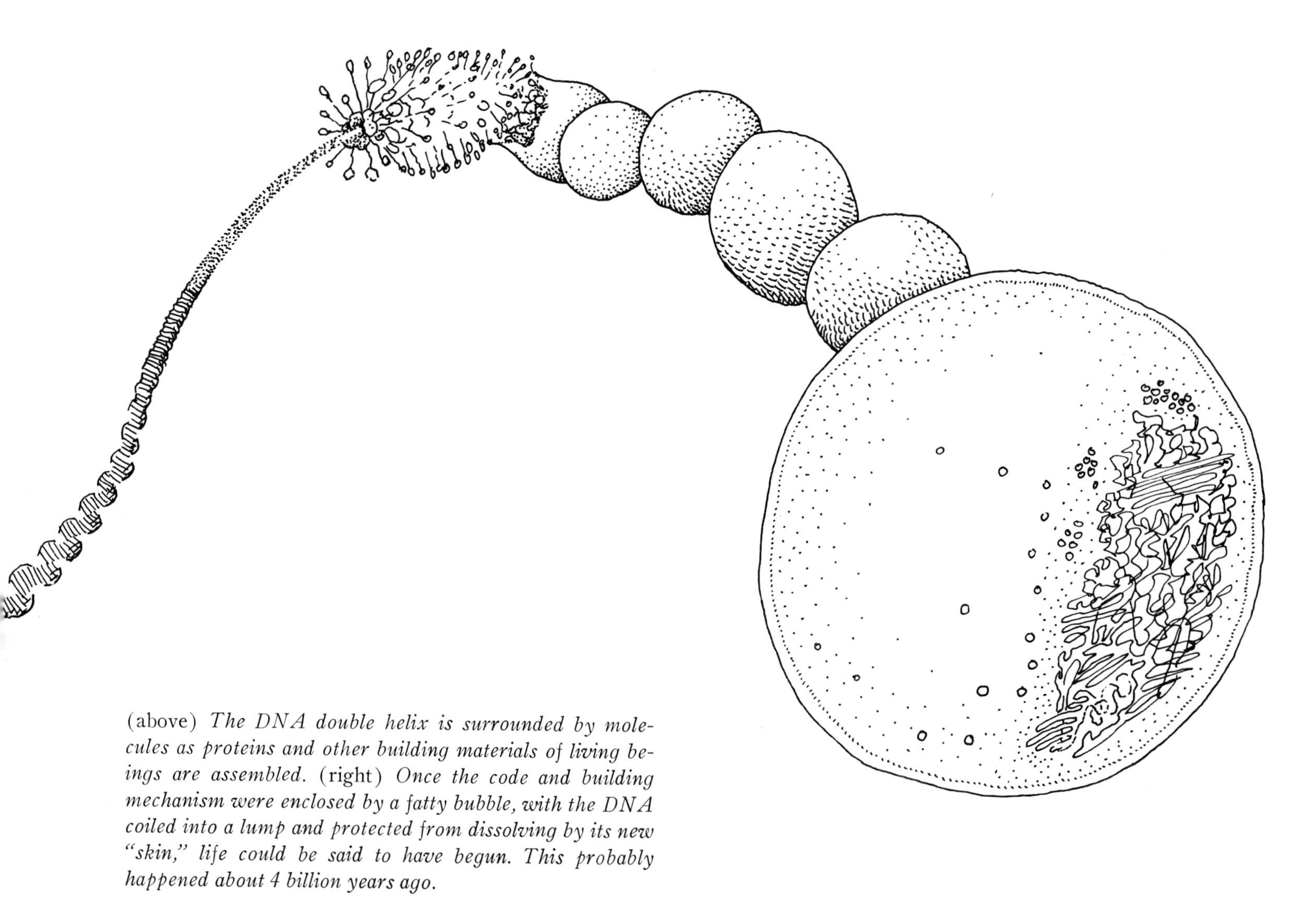

(above) *The DNA double helix is surrounded by molecules as proteins and other building materials of living beings are assembled.* (right) *Once the code and building mechanism were enclosed by a fatty bubble, with the DNA coiled into a lump and protected from dissolving by its new "skin," life could be said to have begun. This probably happened about 4 billion years ago.*

Somewhere along about this time, a mistake—or mutation—occurred that enclosed a bit of concentrated solution of self-replicating carbon-based molecules within a fatty bubble of the oceanside scum. We may never know how this happened, but once it did, the molecules within the bubble, or membrane, had great advantages over those other molecules at the mercy of dissolution in the waters at large.

This membrane ensured the molecules within it enough concentration so that they might always crystallize in times of evaporation. Like a soap film, it allowed the one-way passage of materials from the outer solution when these were in high concentration. This in turn caused water to flow inward through the bubble because of the force of osmosis, and in the beginning such bubble-beings were highly susceptible to bursting. Mutations occurred again, however, and some mechanism appeared (as yet imperfectly understood), which permitted the expanding bubble not to burst, but to divide into two smaller bubbles, each containing a complement of dissolved molecules.

Within the bubbles, mixing went on as always, and there was a gradual increase in the complexity of the enclosed soupy solution. Proteins, vastly complex molecules that order and move the life processes, became a major component of the internal mix, serving as "tools" through which the DNA molecules might more efficiently interact with their unpredictable environment. So began the first real living forms, ever-changing bubbles filled with swirling soups, which depended for their energy on the dissolved chemicals perpetually mixed by energy from the sun.

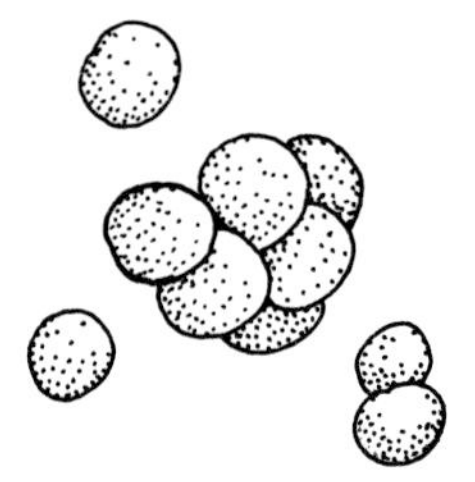

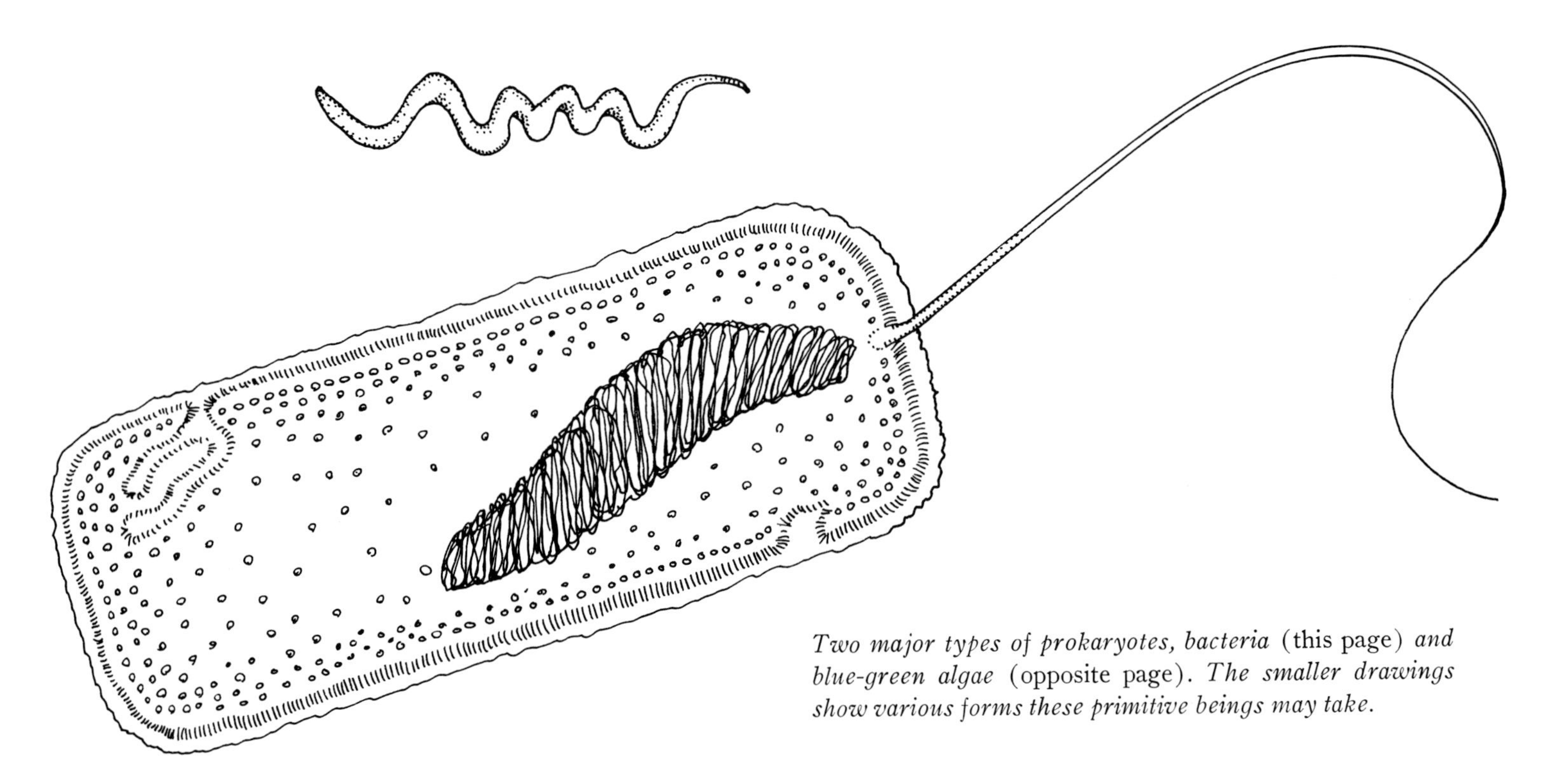

Two major types of prokaryotes, bacteria (this page) *and blue-green algae* (opposite page). *The smaller drawings show various forms these primitive beings may take.*

Very quickly after this stage, there arose two different kinds of beings. One, adapted to the surface of the water, tended to trap sunlight within its membrane in the form of energy-rich molecules of sugar. The other, better able to live deeper in the earth's waters, depended for its food on molecules dissolved in the water. The first of these, which used sunlight directly, we call blue-green algae. Blue-green algae were the first plants. The second group is called bacteria. Together, these primitive beings are called prokaryotes, "before-cells." Their structures are so simple that they are very much different from all later forms of life on earth.

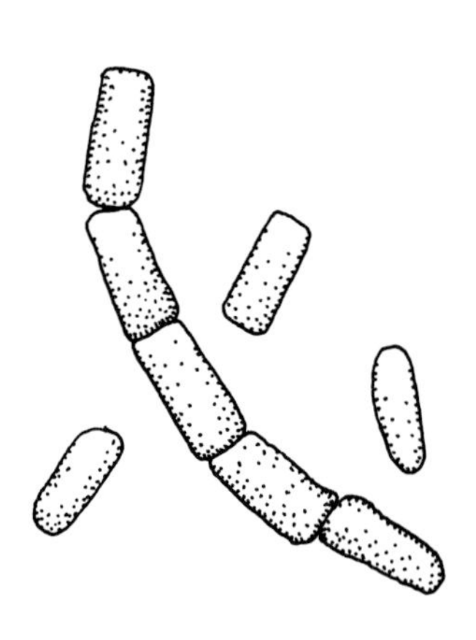

For three billion years, the only living things on earth were bacteria and algae. The algae trapped sunlight and made food out of it, and the bacteria ate that food when algae died and their contents dissolved in the water. Now, it happened that in the process of making food from the sun (called photosynthesis, "light-building"), the algae released into the water a substance we know as oxygen. At that time the earth's atmosphere contained no free oxygen; it was composed of methane, ammonia, carbon dioxide, and other comparatively stable substances. Oxygen, however, is a very unstable substance. Indeed, it is caustic. It eats into steel, devours rocks, and, in those days of primitive life, it was deadly to living things.

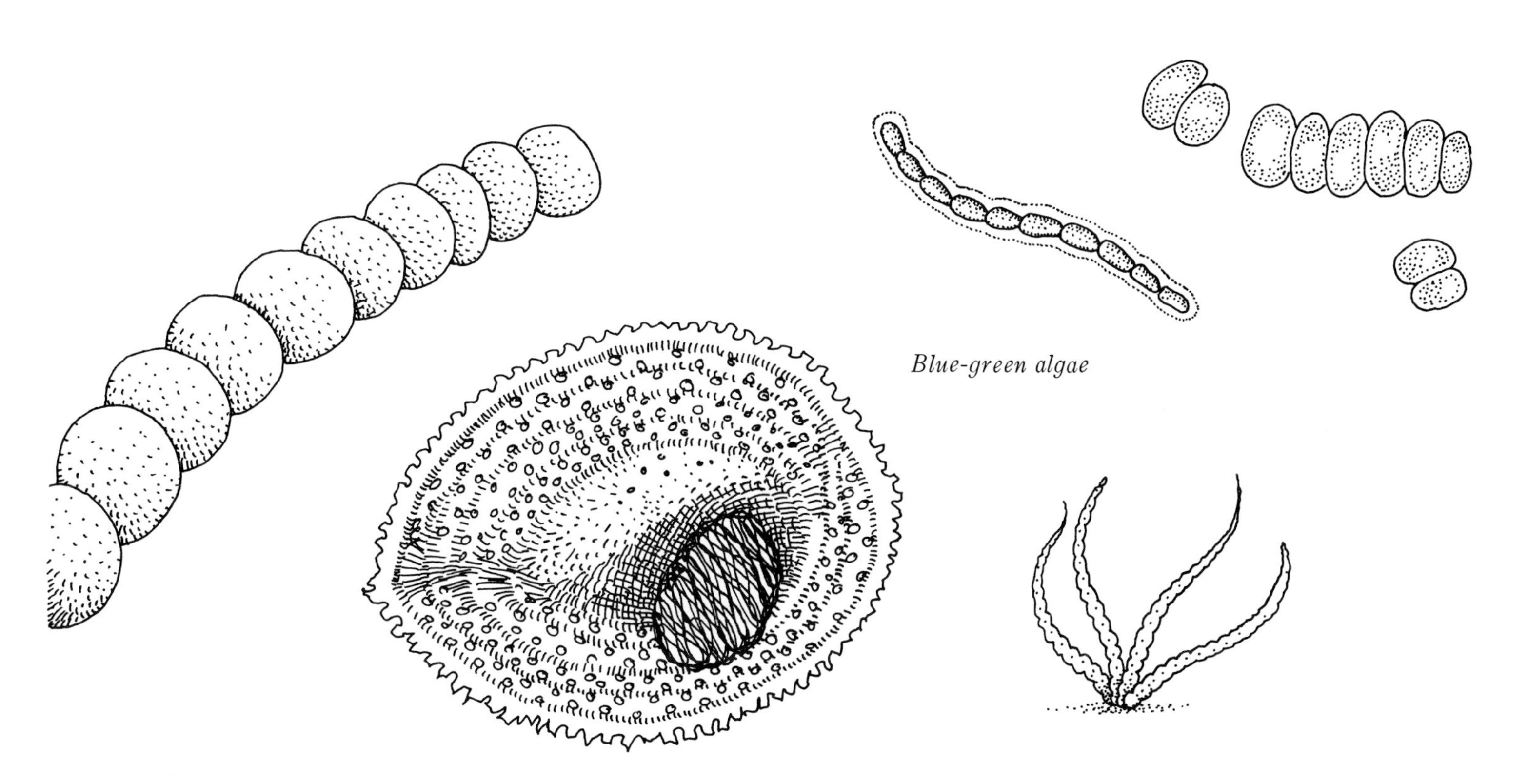

Blue-green algae

Still, the oxygen was released into the water by the algae, and from there it entered the atmosphere above. Gradually its concentration built up until at some point, about a billion years ago, oxygen became so plentiful that living things began to die off, poisoned. Here was an early selective pressure, one that grew in intensity as the years passed. As the pressure increased, those bacteria and algae least able to withstand oxygen poisoning became extinct. Some others, bacteria in particular, took up living in the darker places of the world where oxygen never came. These kinds of bacteria survive to this day, and we call them anaerobes, "those that live without air."

In some bacteria mutations occurred that enabled them to internalize and to use the powerful reactions of oxygen with other substances. These were the first oxygen metabolizers, and they were a great success. We call these bacteria aerobic, or air-loving. Mutations among the algae also allowed some of them to survive the poisoning, and along with the new bacteria, they thrived in the rising oxygen richness of the earth.

Most importantly to our tale, partnerships seem to have arisen in which certain oxygen-metabolizing bacteria shared their energy-using talents with other bacteria and with algae. This sharing, called symbiosis, or "living together," permitted the partners greater options in dealing with their world. It resulted in the rise of what we call eukaryotes, or "true cells." They are tiny beings whose reproductive material is enclosed in a nucleus, or "subcell," within the larger cell. Three grand kingdoms arose from the new symbiosis:

Prokaryotes

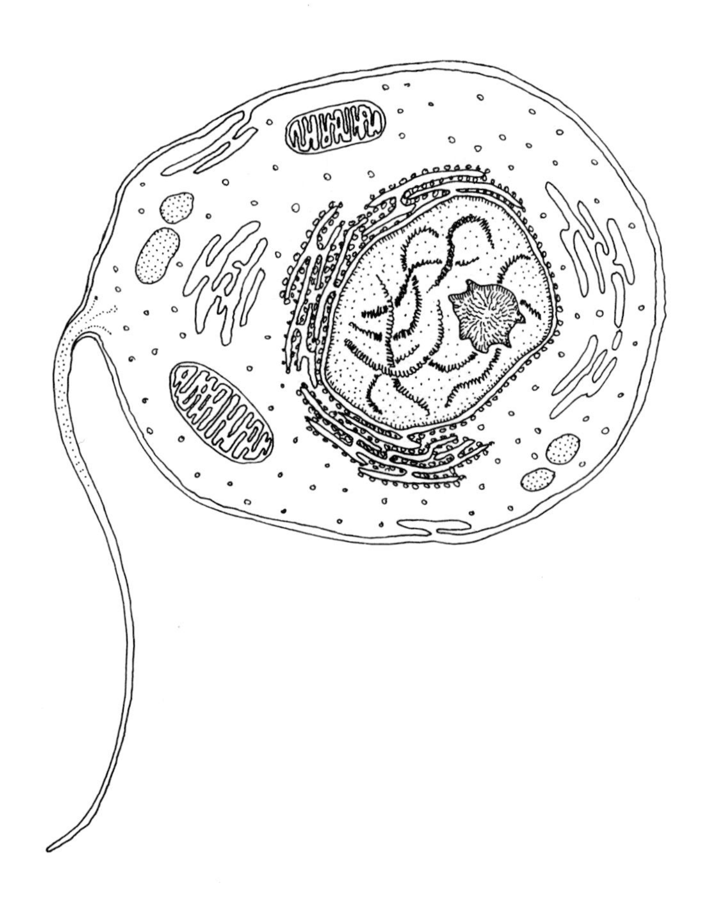

A eukaryotic cell, showing the nucleus within which its chromosomes are protected and ordered. Smaller drawings show (top) *a mitochondrion, an organelle within all eukaryotic cells that processes oxygen;* (middle) *chromosomes, in which the cell's genetic information is stored;* (bottom) *the nucleosome, the "director" of the cell's reproductive functions. Because mitochondria seem to have their own special DNA, it is thought they may be descended from separate prokaryotes that joined a symbiosis about a billion years ago.*

the plant kingdom of photosynthesizing eukaryotes; the fungus kingdom of scavenging eukaryotes; and the animal kingdom of eukaryotes that actively sought and devoured living food. All were powered ultimately by the sun, and all shared common ancestry among the early prokaryotes.

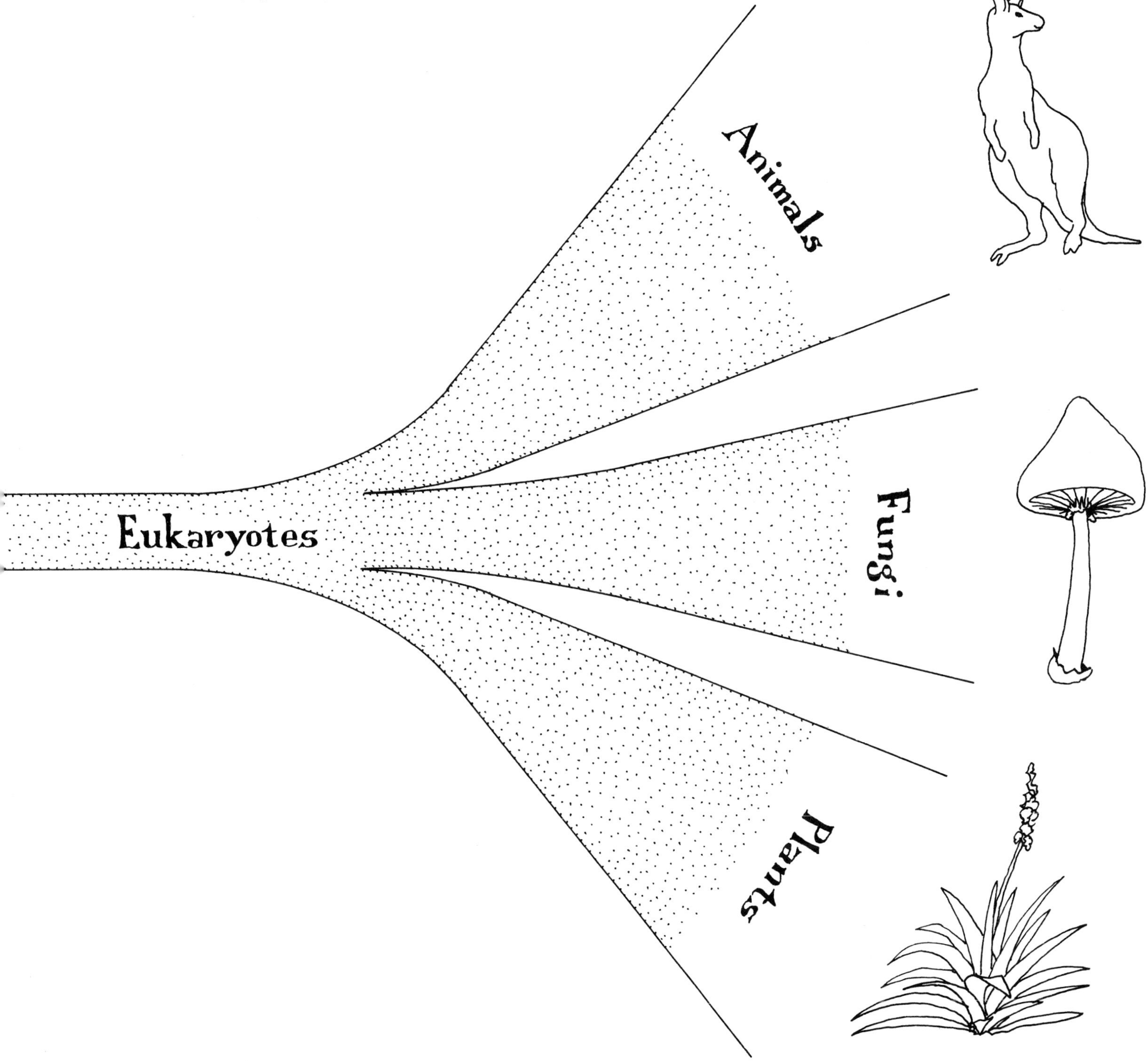

The first adaptive radiation of the eukaryotes produced true plants (autotrophs—feeders on sunlight), fungi (heterotrophs feeding on dissolved foods), and animals (heterotrophs feeding on anything available, and generally characterized by active and purposeful movement).

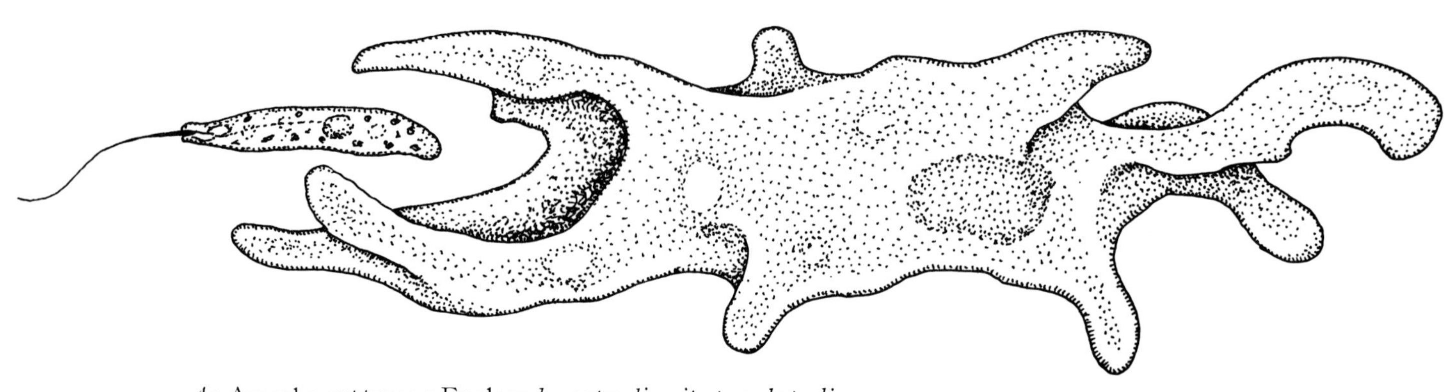

An Amoeba *captures a* Euglena *by extending its pseudopodia.*

Although no clear fossils survive from a billion years ago to show us what the first animals were like, we can get some idea how they may have lived by examining the tiny protozoans—"first animals"—living today wherever there is moisture on the earth's surface. Here we find single-celled creatures competing for resources, much as their eukaryote ancestors must have. During the past billion years, those ancestral eukaryotes evolved into a host of wonderful forms.

All protozoans depend to a greater or lesser degree on movement to get about and live their lives. For this they employ various flagella ("little whips," or propeller hairs), cilia (rapidly waving bands of smaller hairs), and pseudopodia ("false feet," or jellylike extensions of the cell membrane and its contents). Many early protozoans fed on the primitive plants of the time (single-celled algae). The sturdier, harder-to-eat plants survived this selective pressure, and they in turn forced changes among the early protozoans.

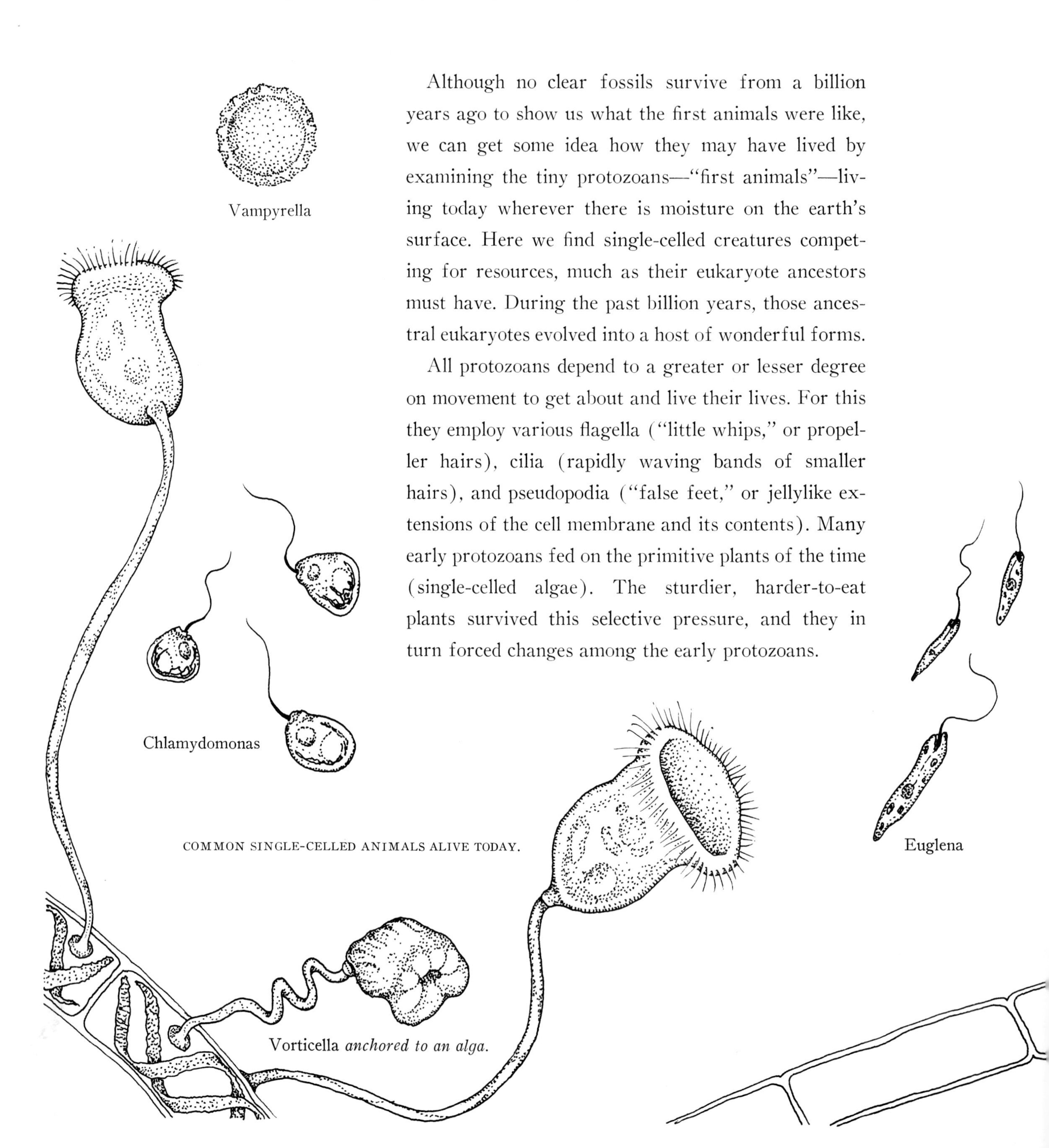

COMMON SINGLE-CELLED ANIMALS ALIVE TODAY.

Vorticella *anchored to an alga.*

Predation, the eating of animal by animal (in this case, protozoans eating other protozoans), became an important way of reusing matter and energy in the early animal world. Predators had a natural selective effect on their prey populations. Protozoans would get eaten or not, depending on the defenses they happened to have. Those with mutations that permitted them greater speed, any sort of armor, or better sensory equipment tended to survive and leave offspring with the same defenses. As a population of prey animals grew faster or hardier or somehow more elusive, it exerted a selective pressure on the predators that chased it: only the fastest or most efficient predators would survive. Over time, a population of predators would become specialized for catching and eating its preferred prey. This perpetually self-reinforcing process, during which two or more forms change in tandem, is called coevolution, "evolving together." It is one of the central processes in the growth of our tree.

Tokophrya *resting on an algal strand spears its prey with tubes that also serve to carry nutritious body fluids from prey to predator.*

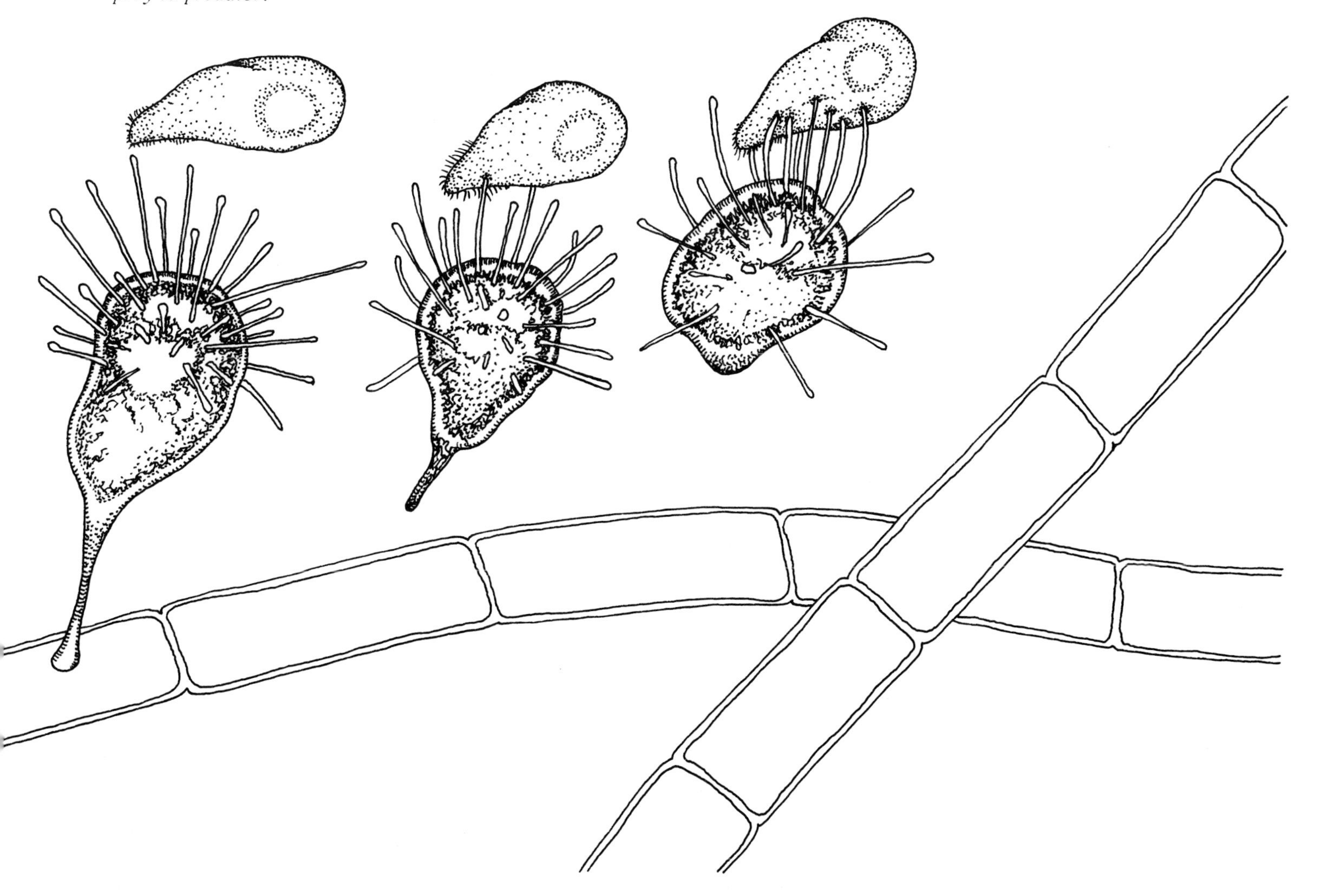

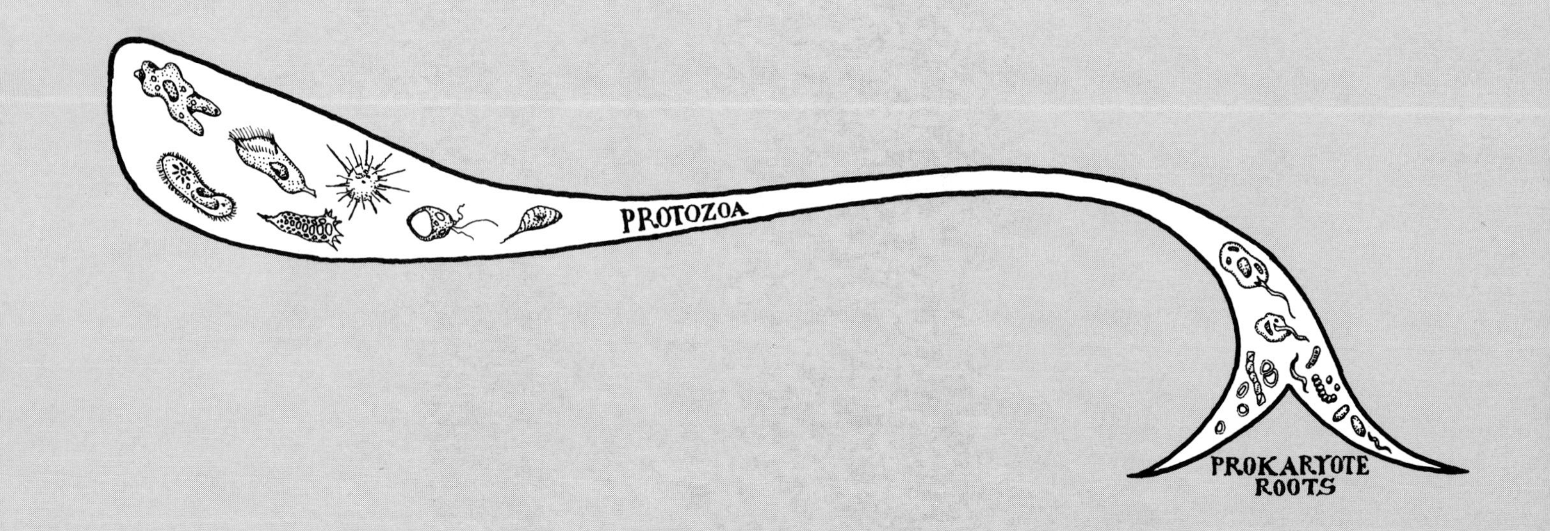
PROTOZOA
PROKARYOTE
ROOTS

There are about ten million living species, not counting many millions more that have become extinct. Since there is not space to show all these minute divisions, the tree will be represented by its larger branches, the phyla and classes.

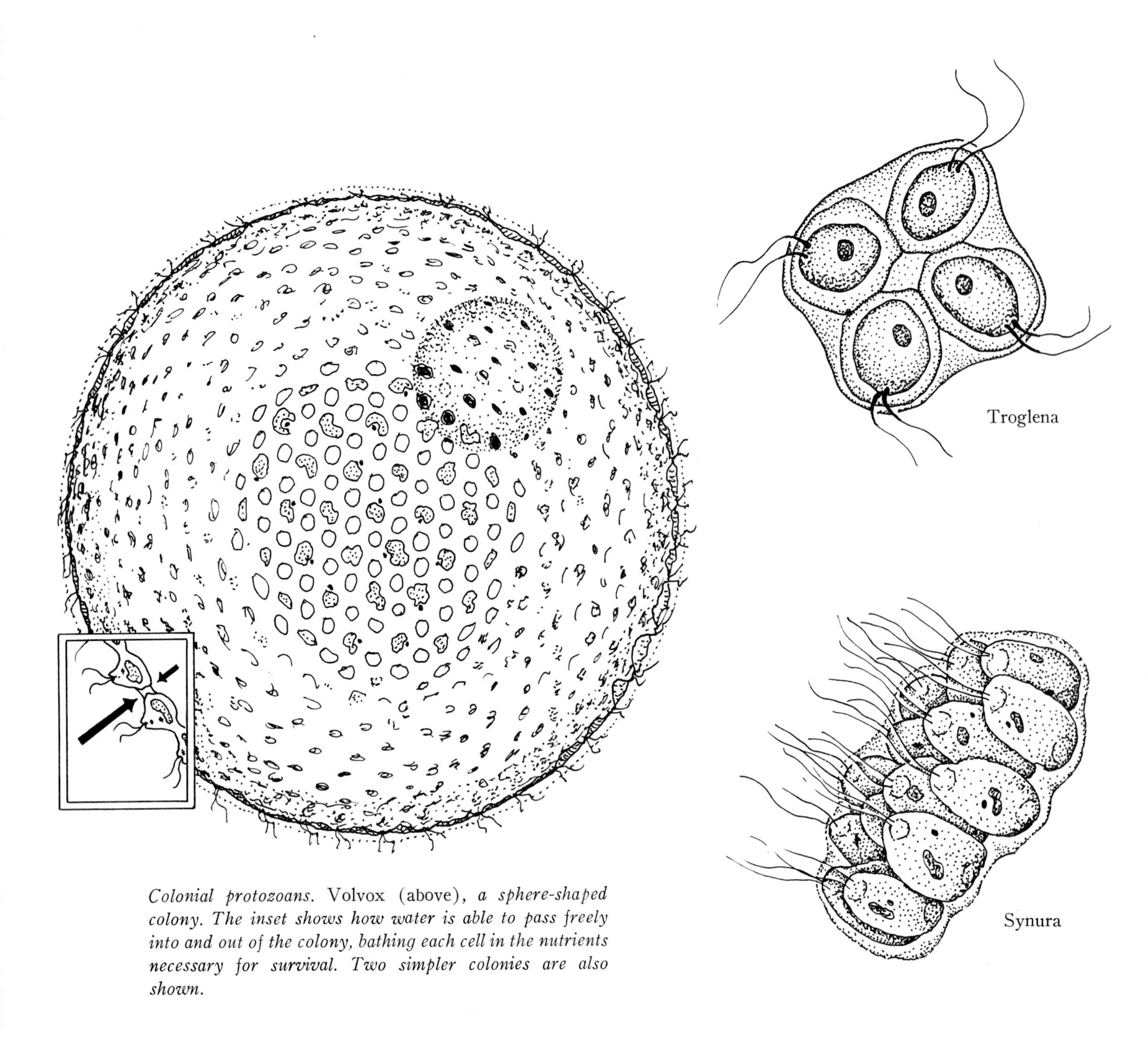

Colonial protozoans. Volvox (above), *a sphere-shaped colony. The inset shows how water is able to pass freely into and out of the colony, bathing each cell in the nutrients necessary for survival. Two simpler colonies are also shown.*

Protozoans continued evolving for hundreds of millions of years (they're still at it), exerting selective pressure on each other and on the primitive plants they consumed. Some of these one-celled beings, perhaps affected by scarce resources, seem to have congregated where moisture and nutrients were most readily available. Gradually these chance congregations assumed the form of orderly clusters. Since protozoans must gather their food and oxygen from the surrounding water, each one in a cluster had to be at least partly exposed to the outside of the cluster. So the first clusters of colonial protozoans were always very thin, about a cell or two thick.

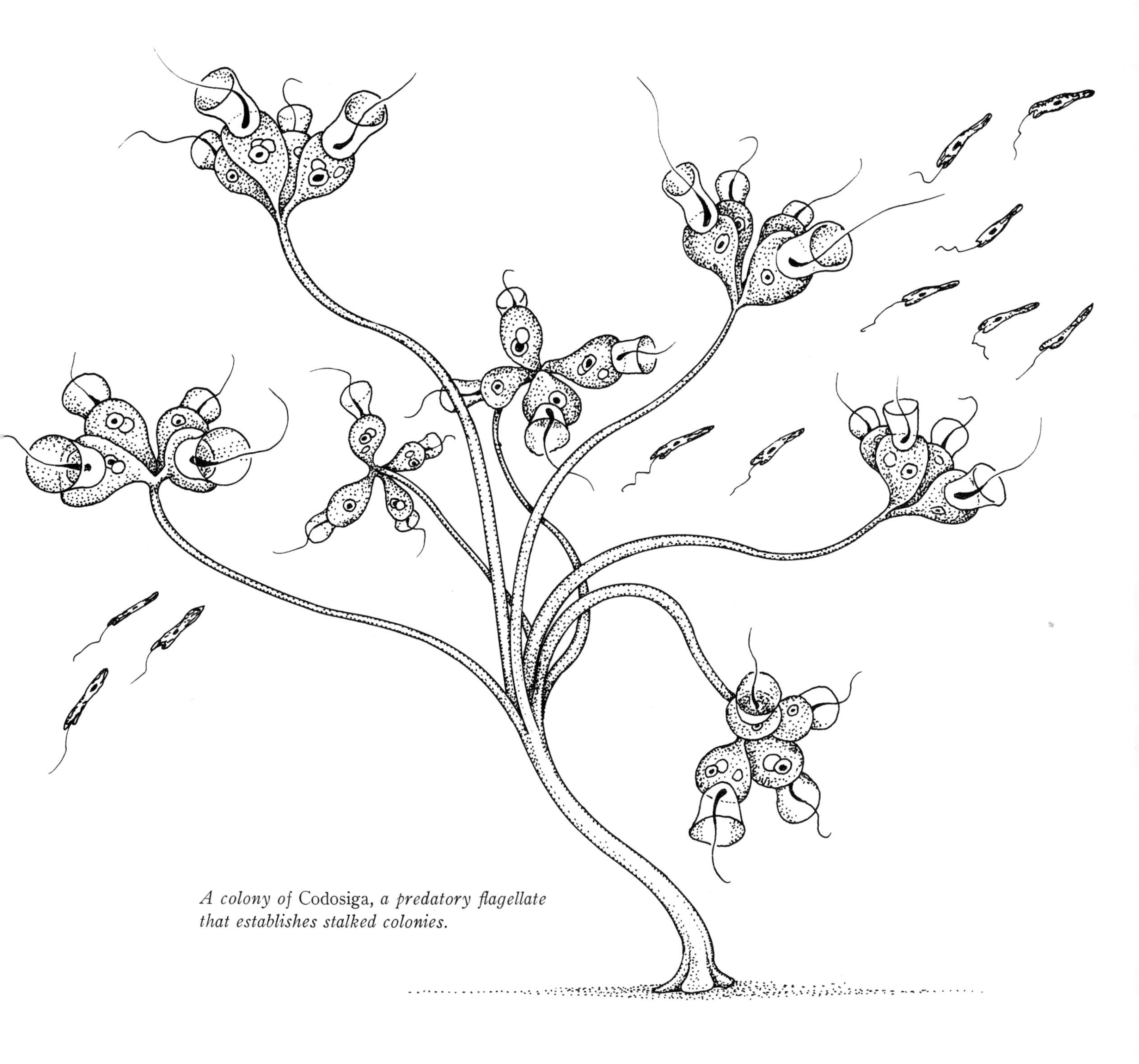

A colony of Codosiga, *a predatory flagellate that establishes stalked colonies.*

There were certain novel benefits for those protozoans that took to clustering. The members of such groups were always concentrated near food sources or other favorable conditions, and in clusters they were less vulnerable to their single-celled predators. Because clustering helped secure nourishment and protection, social protozoans early experienced an adaptive radiation. Colonies appeared in various forms, such as double-walled sheets and spheres enclosing water-filled cavities.

As colonies of protozoans came into competition with each other, those most efficient in distributing the flow of water, oxygen, and food among their members tended to survive. Gradually this competition for resources resulted in the specialization of colony members for certain tasks. Some, by beating flagella, moved water among the members. Others secreted protective armor around the colony exterior. Specialization at this level may be seen today in the Parazoa ("almost animals"), the sponges, which some zoologists still regard as colonial protozoans.

If we separate a sponge into its individual cells, they will rejoin to produce another sponge. Sponges are huge assemblages of cells, most of which are equipped with cilia or flagella that propel water through the sponge's many chambers and tunnels. Some of the cells secrete tiny stiffening structures of silica or calcium, which provide support for the colony. Such colonies may grow several meters across while still remaining simple in basic organization.

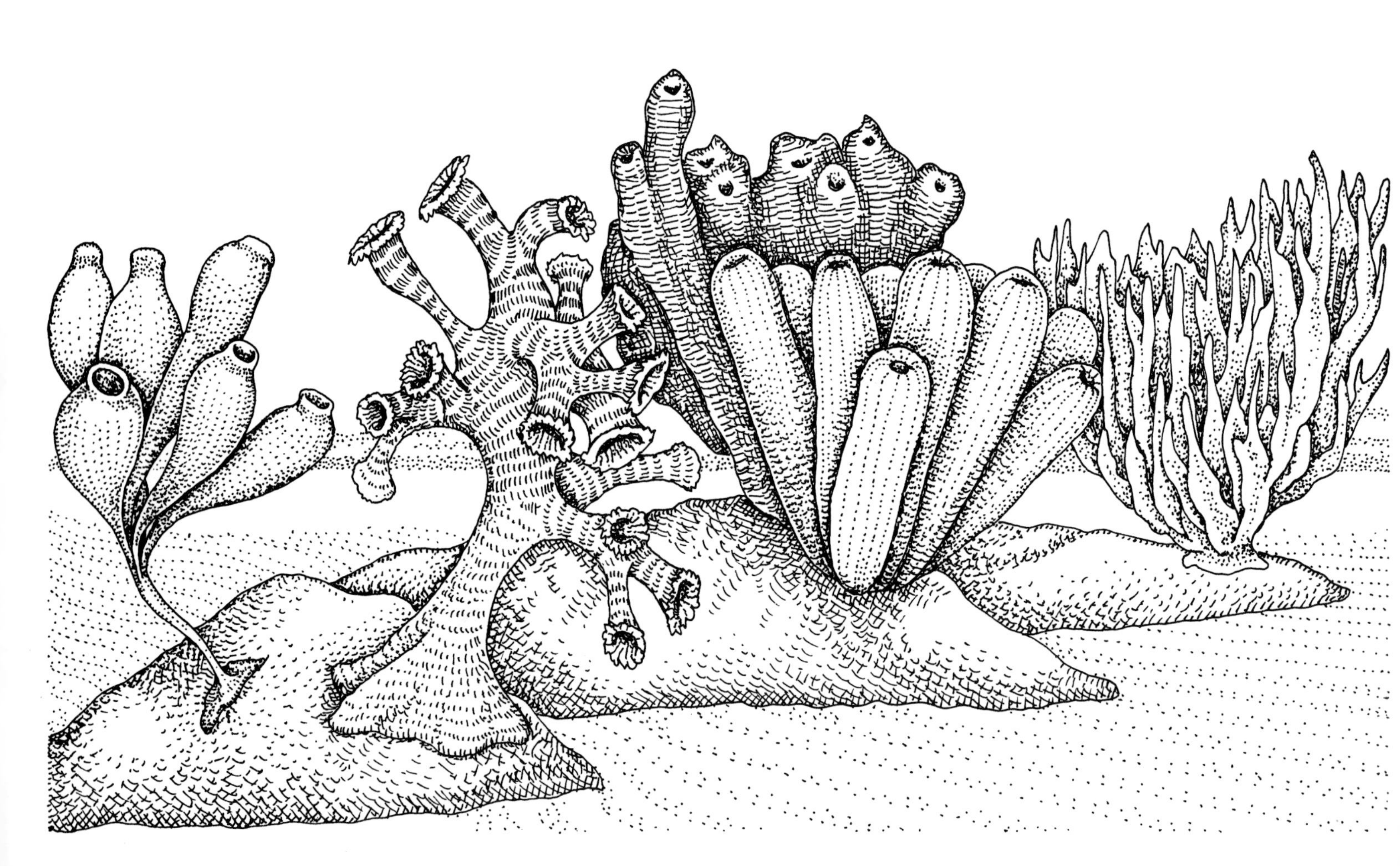

Sponges. A sampling of the many forms these primitive assemblages of cells may take.

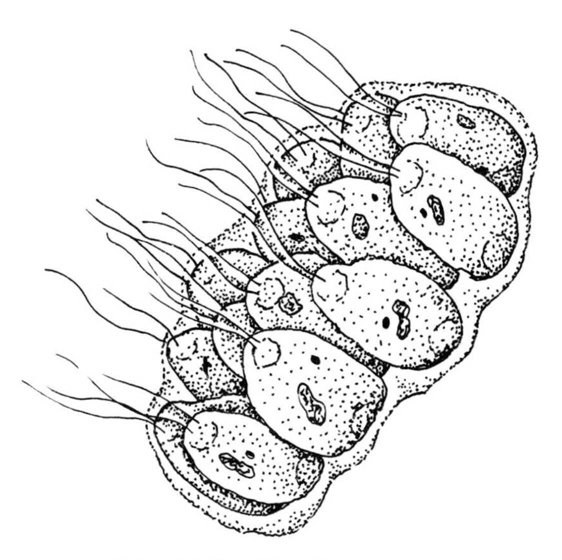

Colonial flagellate Synura

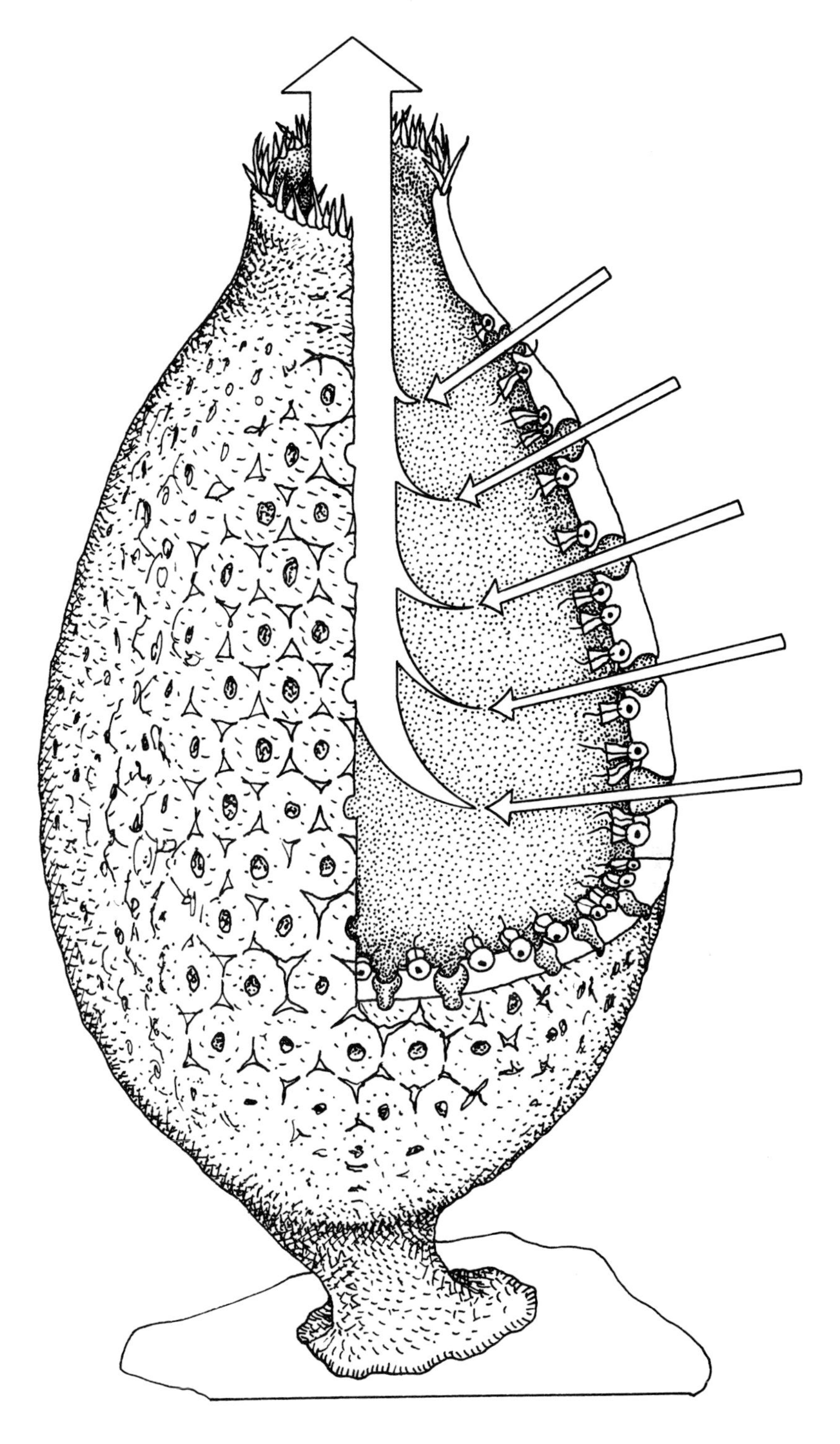

Cross-section of a sponge. Note the similarity between the cells of the sponge and those of colonial flagellates. Sponges cannot be considered true animals; they are more like highly organized colonies of flagellated protozoans.

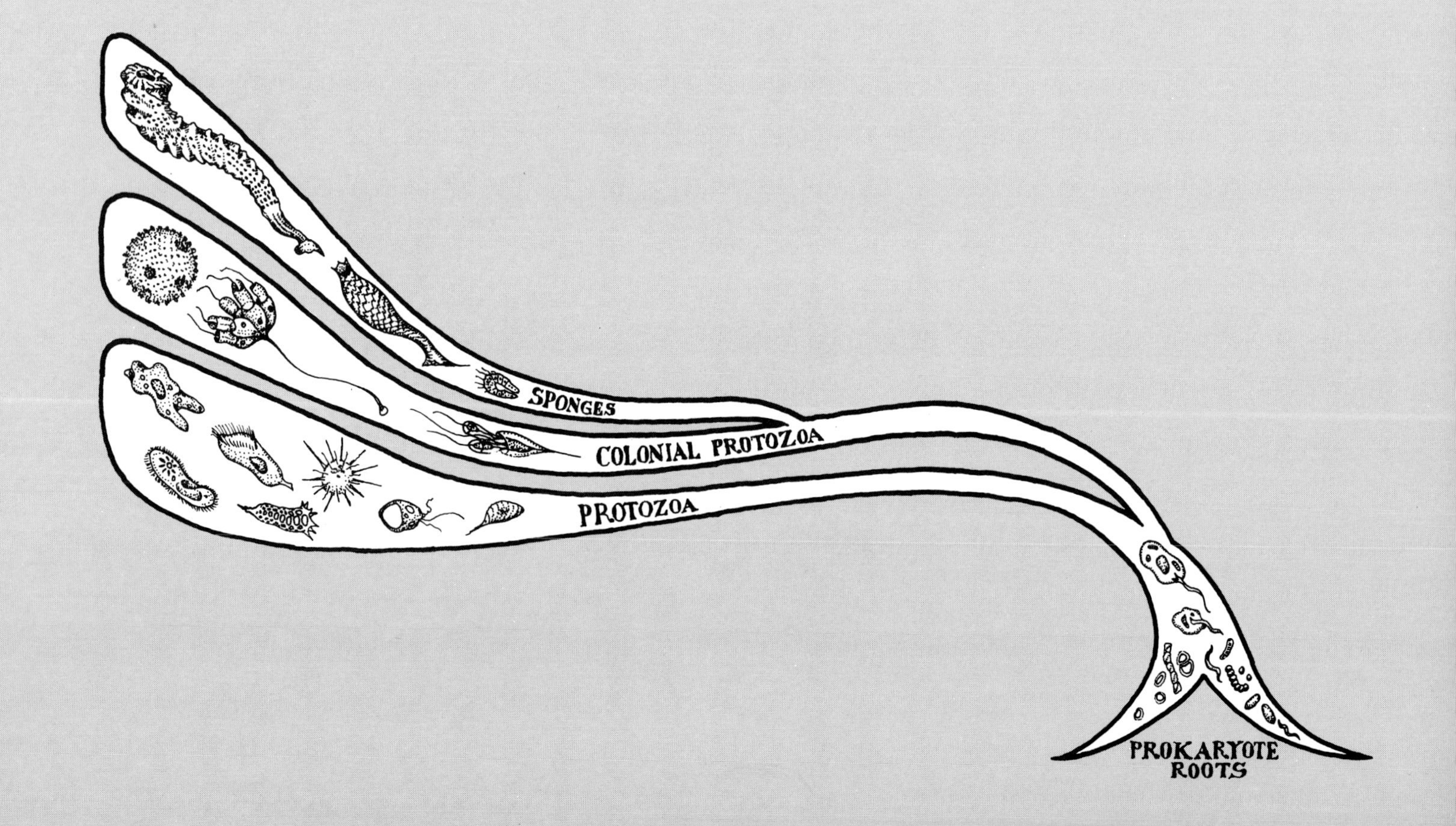
SPONGES
COLONIAL PROTOZOA
PROTOZOA
PROKARYOTE ROOTS

Here we might mention something about the reproduction of protozoans. In good times, given plenty of food, these little animals tend to grow until they reach a certain size, at which point they divide in two. Each of the two "daughter" cells then grows and divides again when it reaches maximum size. For most of the history of life on earth, dividing was the only method of reproduction. Indeed, division is simply an elaboration of the original self-replication of the DNA molecules at the beginning of life some four billion years ago.

When a cell divides, its daughter cells are essentially identical to the parent; they carry the same genetic information as the parent. The mutations that can change the offspring and thus permit evolutionary progress are comparatively rare. There were relatively few kinds of life on earth for the first three billion years—mostly bacteria and blue-green algae, dividers all.

Nevertheless, it happened that at some point in the earth's history, individual organisms appeared that occasionally exchanged genetic material with one another; certain bacteria do this today. This exchange led to a mixing of genes, the "instructions" of heredity. When the cells that traded genetic material reproduced, they each had daughter cells that were in some way different from themselves, due to the addition of genes from the other cell. This was an important early step that speeded the evolutionary process, creating as it did a rapid increase in the frequency of inherited changes.

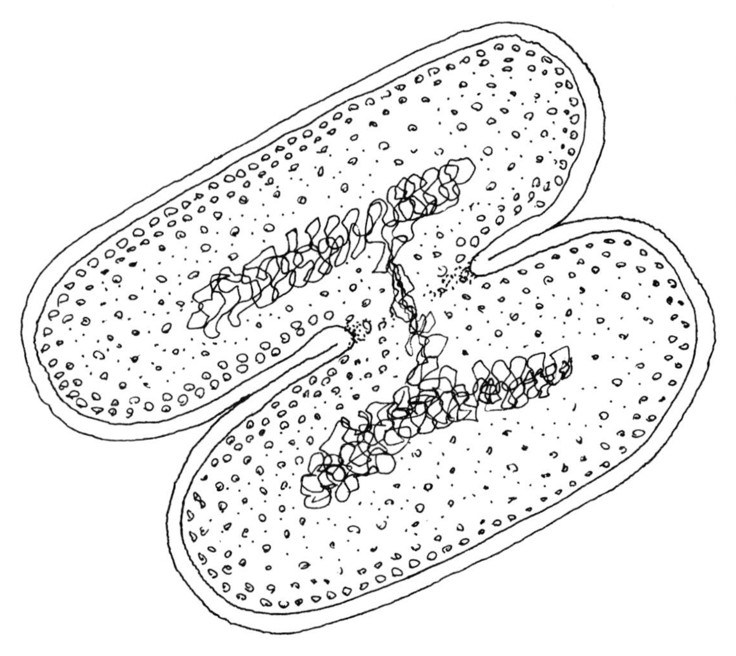

Occasionally some bacteria exchange genetic information before dividing (left) ; *while this is not true sexual reproduction, it is a forerunner of sex, in that it permits "new" individuals to appear rather than simply two identical daughter cells as in simple division* (below).

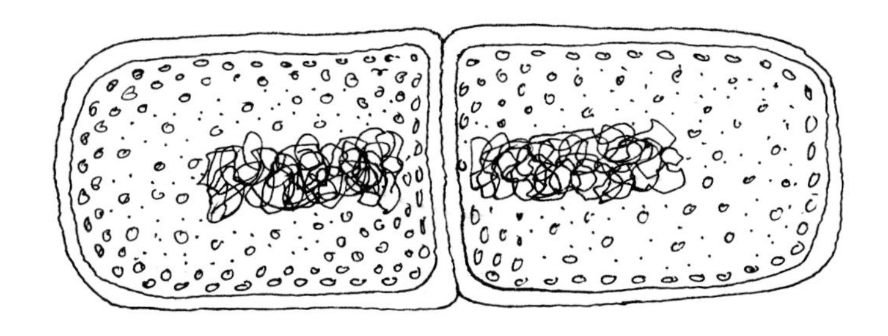

Ultimately true sexuality arose, probably in some early eukaryote whose ancestors practiced this genetic exchange. Within a single species, two kinds of individuals appeared, each with a different method of containing and distributing the hereditary information. We call these two divisions of a species the sexes, and we distinguish them (sometimes this gets hard!) on the basis of which one of the two invests the most energy (food) necessary for the initial growth of the offspring.

The big investor of energy is the female. She creates in her body an egg, in which the initial development of the offspring will take place. The male, on the other hand, does not expend much energy in the reproductive material. In most animals the male offers only a tiny moving cell, which contains half of the information needed to grow a new individual. The egg, the female reproductive cell, contains the other half of the information in addition to the all-important food supply. When these two halves meet, all the ingredients necessary to create a new individual are present, and the process of growth begins.

The male half of the genetic material is contained in some sort of moving vehicle (a sperm, a microscopic flagellated swimmer). The female half, the egg, is comparatively immobile. As an outgrowth of this fundamental character of the sexes, it is noticeable that in almost all but the most complex animals the male seeks out the female rather than vice versa. Indeed, the female, like her sex cells, is usually rather larger than the male, a size difference that reflects the greater investment she makes in the welfare of the offspring.

The advent of sexual reproduction initiated a great mixing of genetic material. It is probably at least in part due to sex that so many different kinds of animals appear in the fossil record at once, for the rate of evolution must have increased in response. Sex, the great evolutionary mixer, was also the great accelerator.

Colonies of protozoans paved the way for the evolution of Metazoa, or "higher animals," the true multicelled animals that include everything from jellyfishes to ourselves. One cannot take a metazoan, break it into separate cells, and expect those cells to rejoin and create that original metazoan! The level of cell specialization has passed too far beyond the point of simple colonialism. When metazoans appeared, the way was open for the evolution of much larger and more complicated animals.

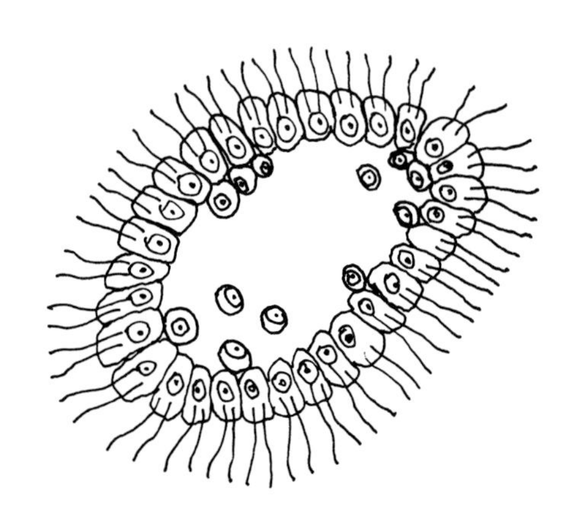

The mobile larva of a sponge; it is very much like a ball of flagellated protozoans.

Metazoans probably originated in the form of small colonies of protozoans that bore flagella by which they moved about. Unlike simpler protozoan colonies, however, these had a definite fore-end and rear end. Such little colonies, called planulae ("little flat ones"), were bullet-shaped swimmers with no real eyes, internal organs, or, in fact, much of anything. They were, however, fast and efficient for their time, and they appear to have given rise to a mighty empire of metazoans perhaps 800 million years ago.

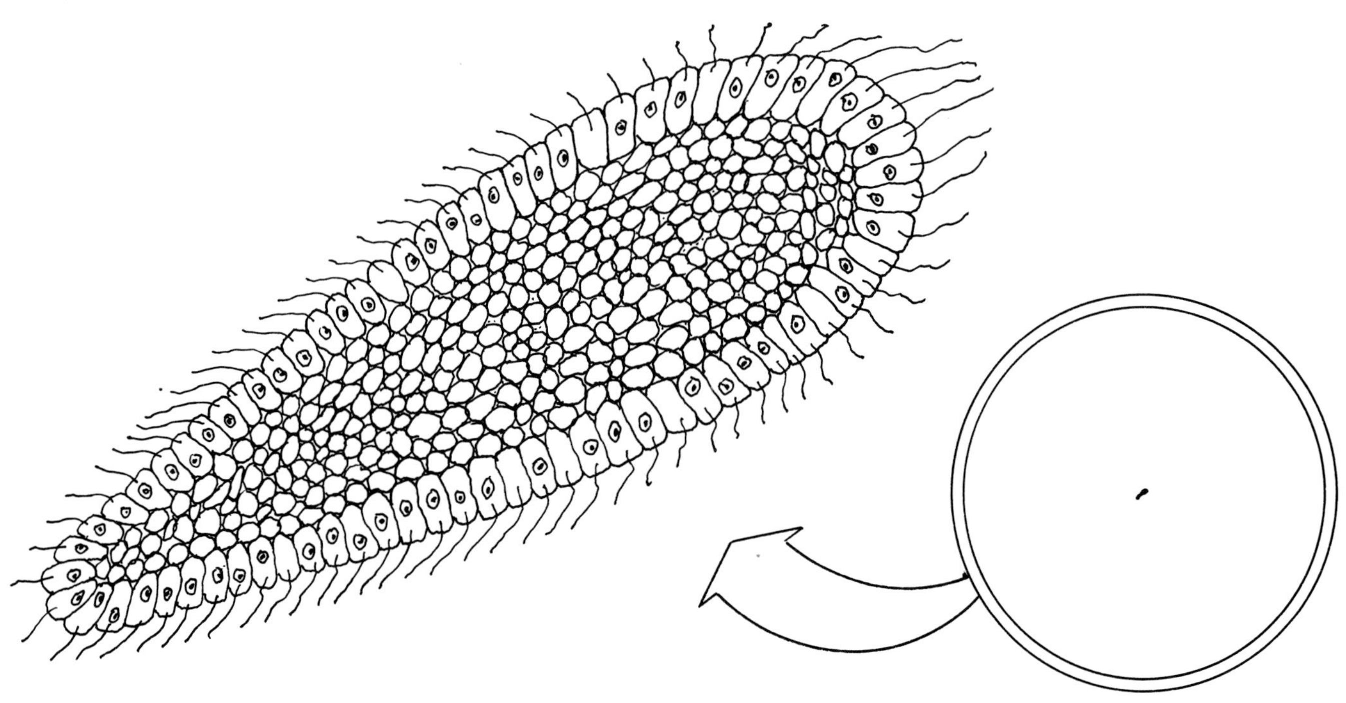

The planula, an assemblage of cells that lies at the evolutionary foundation of all the metazoans, or higher animals. Actual size is shown in circle.

Our first clear look at early metazoans is provided by beautifully preserved fossils from the hills of Ediacara in Australia. But these fossils, captured in fine muds laid down on the bottom of a shallow sea between 600 and 700 million years ago, represent highly organized metazoans. Their planulae ancestors must already have experienced millions of years of evolution as multicelled beings. For the soft, small ancestral metazoans, there is no fossil record.

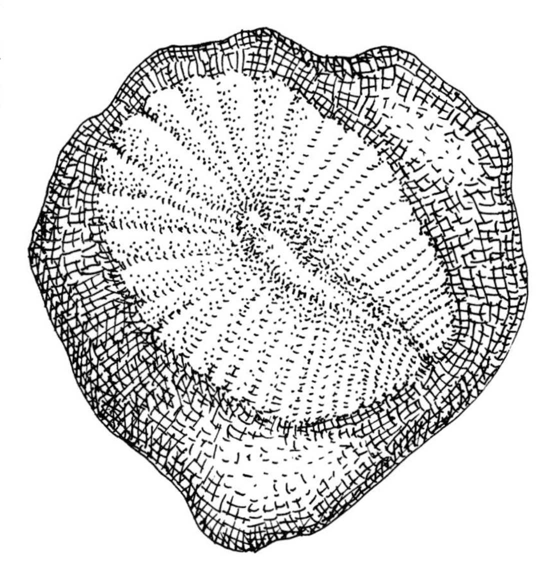

Dickinsonia, *a possible wormlike form*

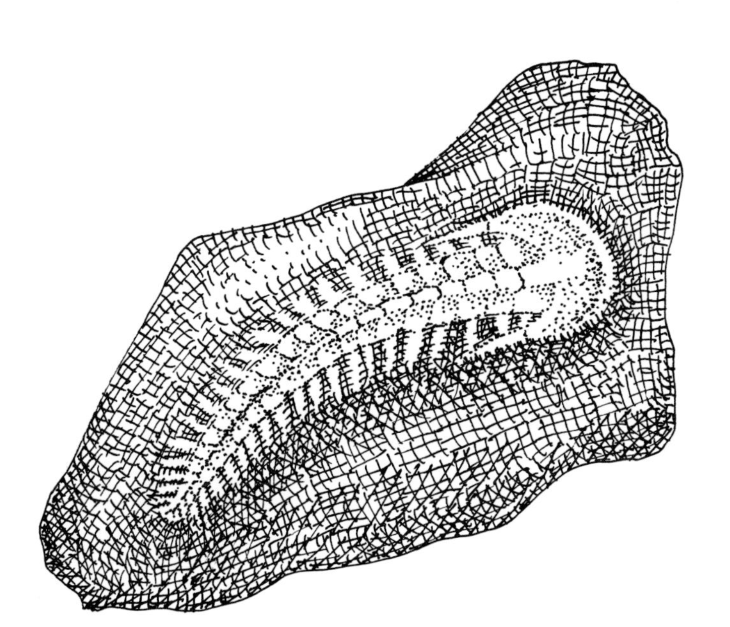

Spriggina, *an early worm*

FOSSILS FROM THE HILLS OF EDIACARA IN AUSTRALIA.

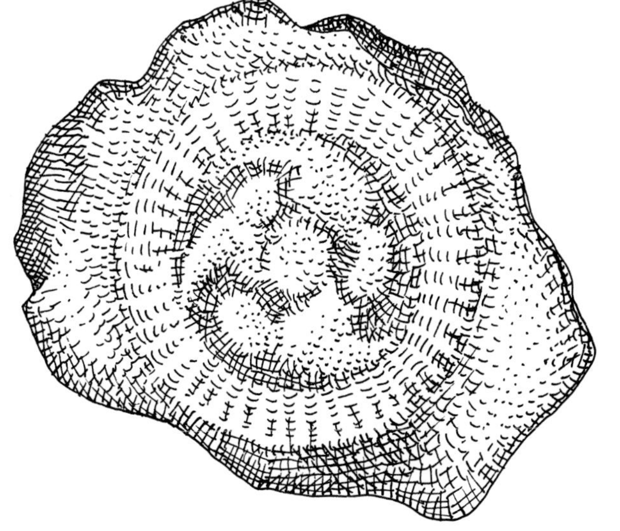

Unidentified (possibly starfish-like) animal, Tribrachidium

The Ediacaran animals were complicated creatures for their time, sporting a variety of adaptations fitting them to a quiet, bottom-dwelling way of life. Most of them were soft-bodied animals among whom paleontologists have found representatives of at least three major groups of animals still living today. There were members of the phylum Coelenterata, "hollow-innards," a group that today includes jellyfishes, corals, hydras, sea anemones, and their relatives. Coelenterate larvae are still close to the old planulae in form. There were representatives of the phylum Annelida, "ringed ones," the

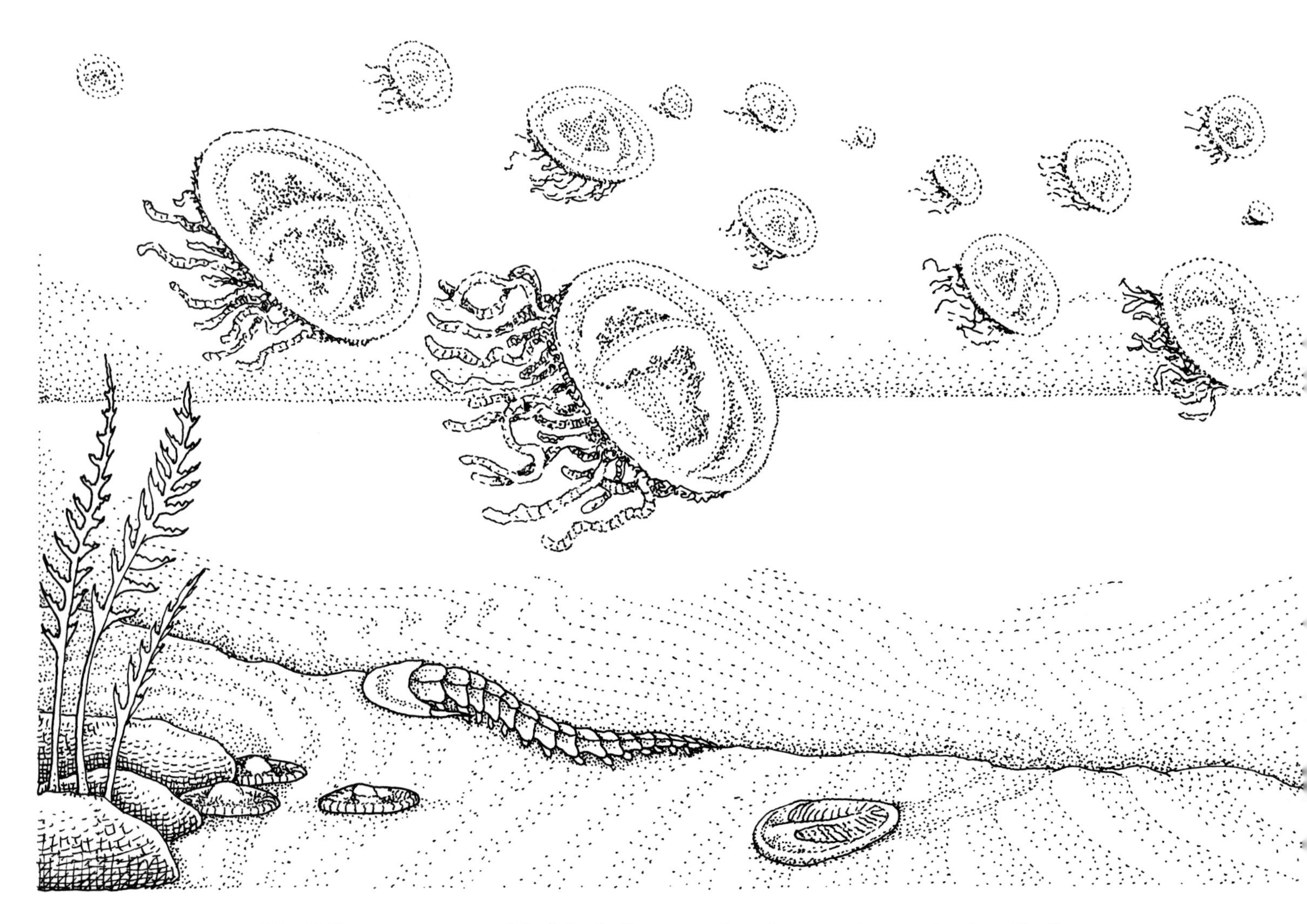

The Ediacaran seascape. Jellyfish similar to modern forms swim overhead, while "sea pens," stationary fern-shaped animals which filter particles of food from the water, grow

advanced phylum of segmented worms, of which the living earthworms are perhaps best known today. (Of course, in Ediacaran times all life was still water-dwelling.) And there were members of the phylum Mollusca, the "soft ones," from whom are descended clams, snails, octopuses, and their relatives. There are other animal fossils in the Ediacaran record, but they do not closely resemble anything alive today and probably represent branches of the tree that were for one reason or another ill-adapted to survive to present times.

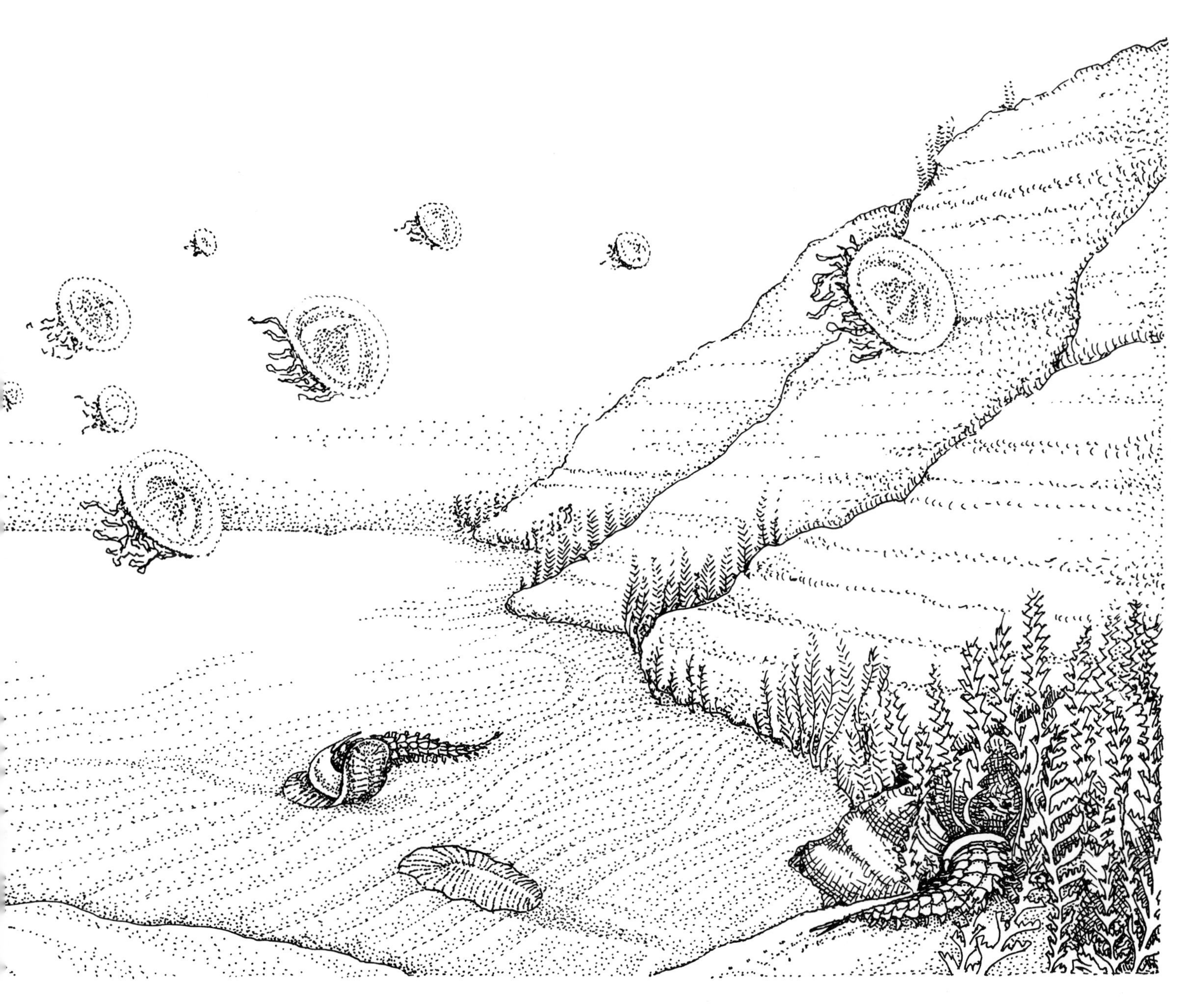

at extreme right and left. On the sandy sea bottom, wormlike crawlers experiment with early models of legs and complex musculature.

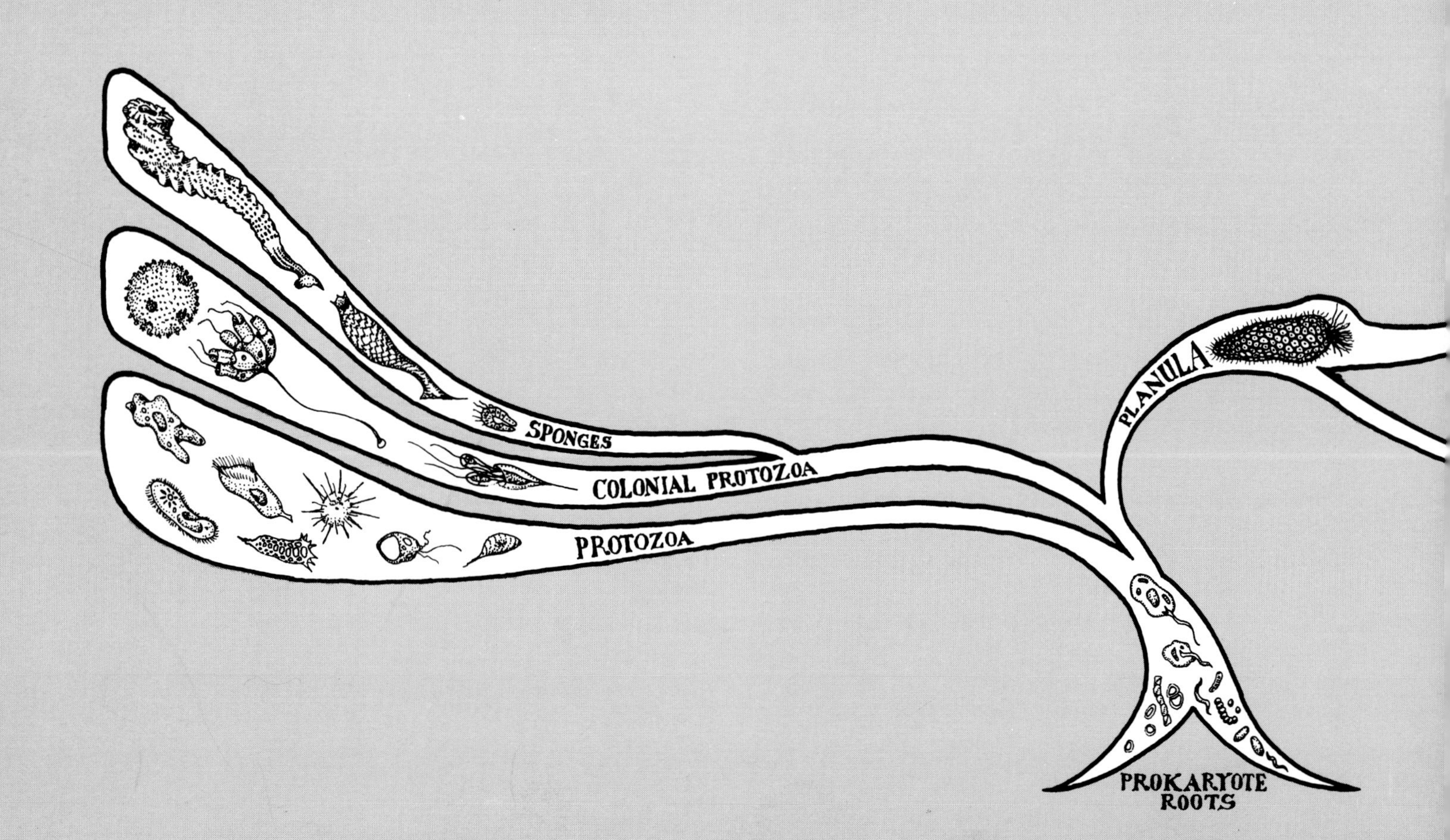
SPONGES
COLONIAL PROTOZOA
PROTOZOA
PLANULA
PROKARYOTE
ROOTS

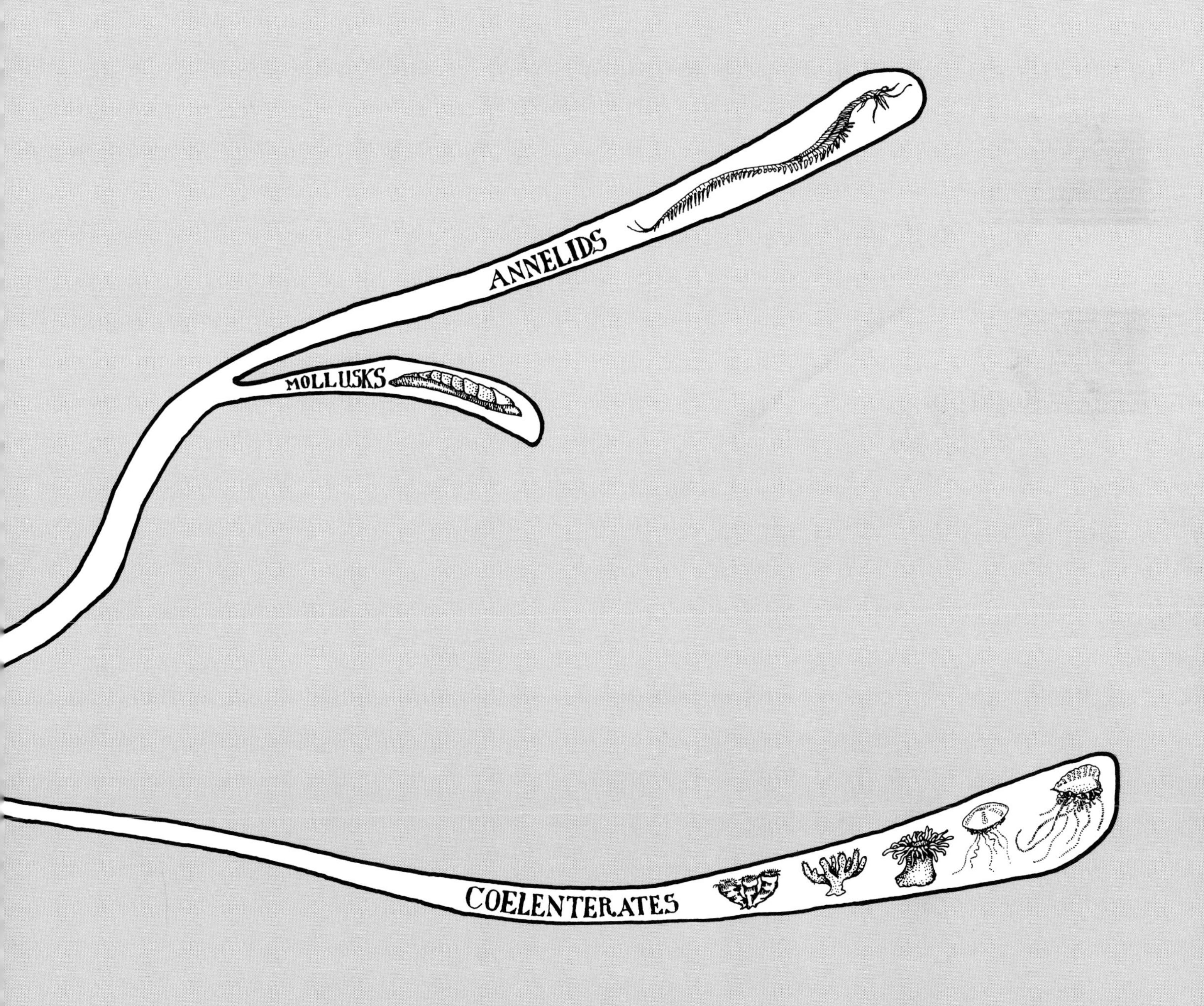
ANNELIDS
MOLLUSKS
COELENTERATES

Judging from their looks, the Ediacaran animals had an easy way of life. They moved slowly, eating algae and preying on the slowest and most vulnerable among them. Gradually, as the weaker Ediacarans died out, the Ediacaran population as a whole grew tougher and more diversified. The prey-Ediacarans acquired defenses against being eaten—they got faster or nastier tasting, for instance—and the predator-Ediacarans were forced to develop in directions that would overcome those defenses. Predatory Ediacarans lacking traits that permitted them to catch faster, tougher prey simply didn't survive.

One of the most important changes resulting from the coevolution between predators and prey was the development of armor of various sorts—the first external skeletons or shells. A spiny, lumpy animal, or one encased within a sturdy container, is harder to eat than a soft, unprotected one. In addition to offering protection, armor also offers physical support to its possessor and permits the animal to grow to larger size. Armored creatures are also better able to move their muscles, because they have a rigid structure against which to push and pull. Coelenterates, for example, are soft and jelly-like. But by Ediacaran times, the first corals had evolved, coelenterates that possessed external skeletons. The most primitive coral polyps lived alone and secreted limy cups in which the polyps themselves sat, like eggs in eggcups. As they evolved further, corals began growing in groups, secreting limy cups that stuck together in masses of skeletons (coral reefs), covering the rocks on which their ancestors had fastened themselves.

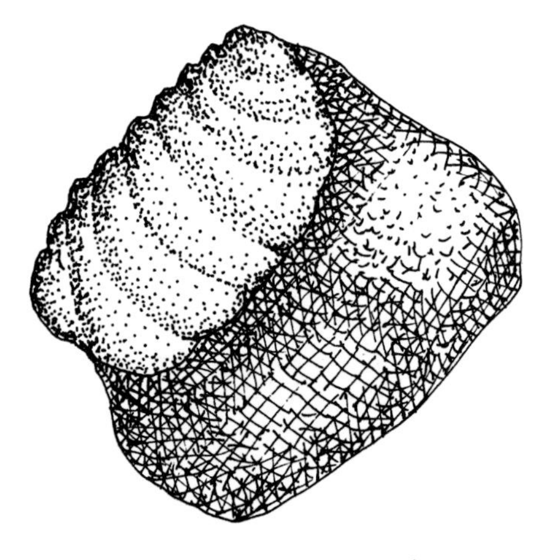

A primitive mollusk, Heliconella

FOSSILS OF ANIMALS FROM THE BURGESS SHALES, LAID DOWN ABOUT 530 MILLION YEARS AGO.

It is usually only the hard skeletal parts of animals that become fossilized, leaving their imprints in the geological record to be read by us, their descendants, many millions of years later. The soft parts of most mollusks, for instance, were encased in shells like those of the familiar clams, snails, or nautiluses. Fossils of these shells provide the major record of this ancient phylum and offer us a look at its evolution through hundreds of millions of years.

By about 530 million years ago, the tree of animal life had blossomed in earnest. The fossils of that period, many of which are found in the Burgess Shales of British Columbia, show an ocean-bottom world populated by a wide variety of animals, including ancestors of many phyla that survive to this day. In addition to the sponges, the annelids, the mollusks, and the coelenterates, there were members of the phylum Echinodermata ("spiny-skinned"), the starfishes and their relatives. Brachiopoda ("arm-feet"), the lampshells, and their close relatives the Bryozoa ("moss animals") had also appeared. These are tiny animals that live in underwater colonies often mistaken for seaweed or corals. There were members of the phylum Arthropoda ("jointed leggers"), first in the form of trilobites; later this phylum would include the familiar crabs, lobsters, spiders, insects, and other such jointed beings.

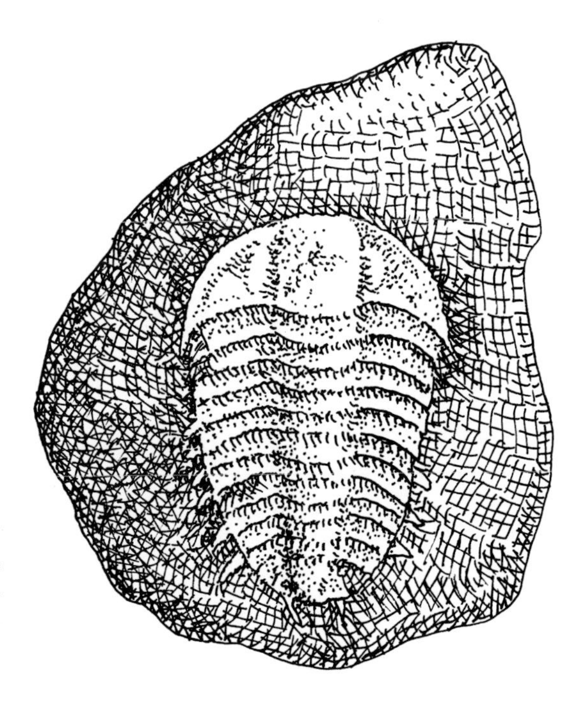

A trilobite, one of the earliest expressions of arthropod life and among the most successful forms ever to have existed. Trilobites ranged from insect size to about fifty centimeters.

There were even the first members of our own phylum, the Chordata, animals with internal stiffening rods against which their muscles might play in swimming. Every chordate also possessed a nerve cord that lay along its back. Later on, the fore-end of this cord would swell into a real brain. But those chordates we find in the Burgess Shales were simple pen-shaped swimmers, very much unlike the colorful and diverse backboned animals that inhabit the earth today.

Although their fossils do not survive to show us exactly when they appeared on the tree, we can imagine that many other groups of soft-bodied animals had come into being by Burgess times. Among these are the roundworms, some of the most numerous animals living today. Although most modern roundworms are parasites in larger plants and animals (the hookworm, pinworm, and the tiny worms that cause trichinosis are familiar examples of roundworms parasitic on human beings), their ancestors were probably swimming or burrowing worms of simple construction that ultimately took refuge as parasites in other animals to escape their more agile neighbors. Similarly absent from the fossil record but probably equally ancient are the flatworms.

A primitive crustacean, relative of modern crabs and lobsters

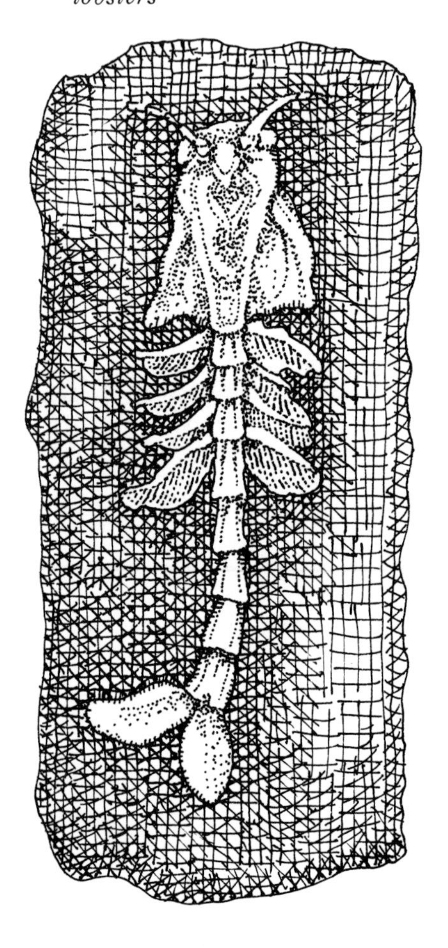

A sponge, Vauxia

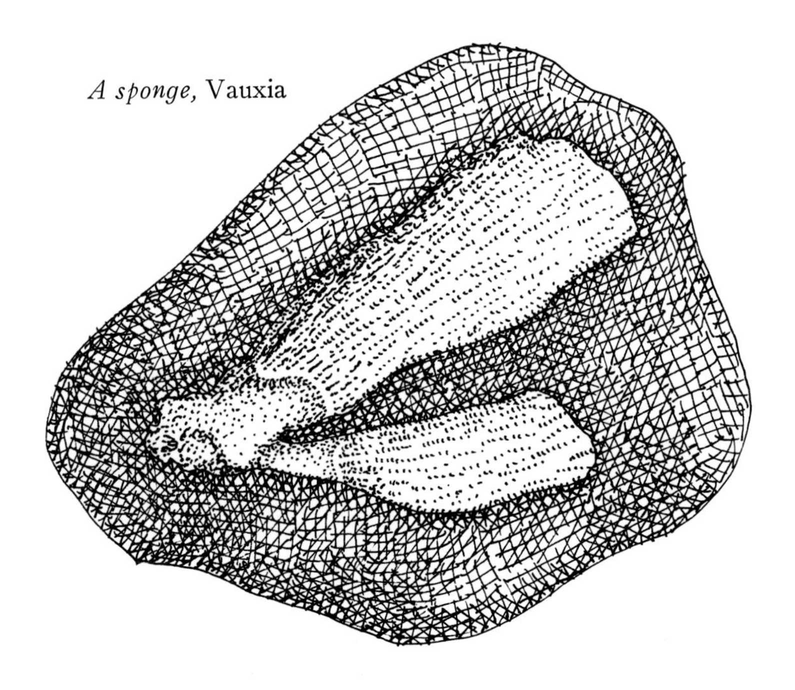

BURGESS FOSSILS

Related to roundworms are rotifers, or "wheel-animalcules." These tiny beings, many as small as protozoans and often mistaken for them, actually have many cells and are fairly complex in form. Their name derives from the ring of cilia, or moving hairs, around their mouths with which they suck in the water containing their microscopic food.

The Burgess sea bottom was a strange world of little chordates, ferocious burrowing and swimming marine worms, corals, jellyfishes, and crinoids ("sea lilies," relatives of starfishes that lived anchored to the sea bottom like long-stemmed animal "flowers"). Its many arthropods included early forms of the very successful tanklike trilobites, which looked and probably lived rather as do the horseshoe crabs of the northern Atlantic Ocean today. Trilobites, in numbers at least, were the dominant sea-floor animals for about 300 million years.

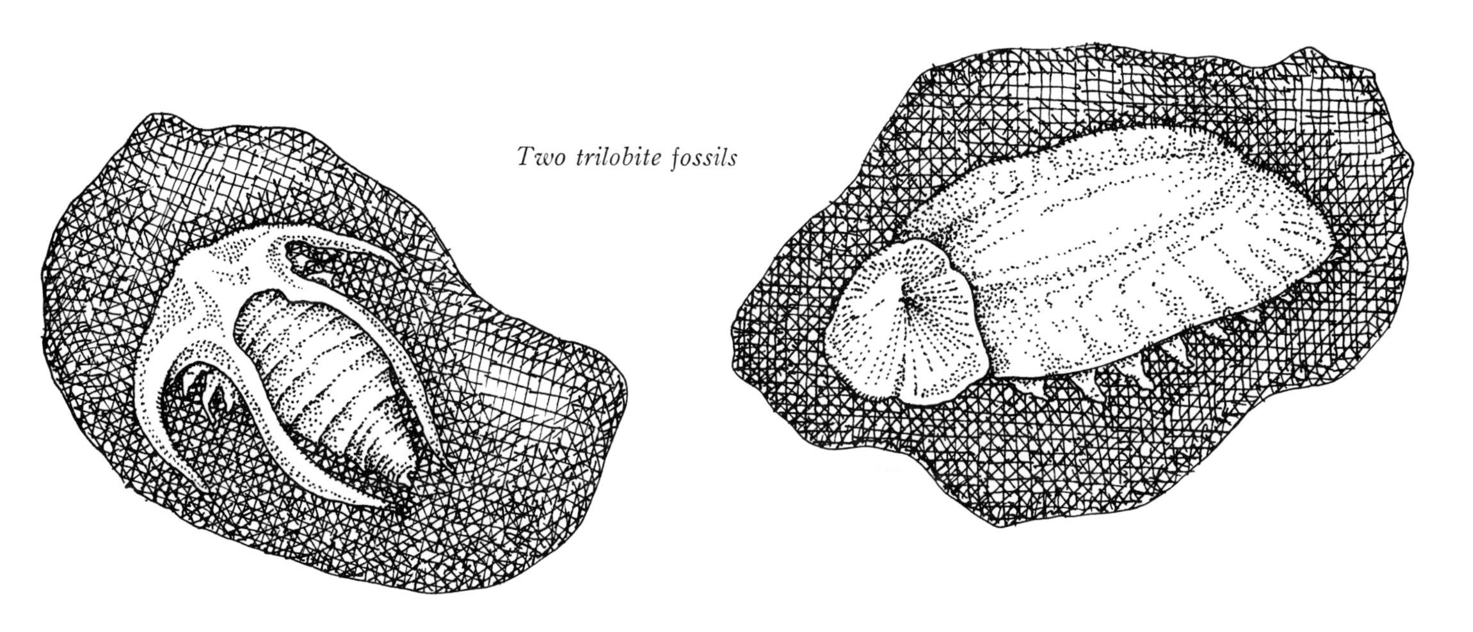

Two trilobite fossils

There were plenty of other animals sharing the Burgess mud, including some the likes of which we cannot find alive today. There were saucerlike five-eyed swimmers that floated along the bottom. There were long segmented beings that snagged their prey with single pincer-tipped tentacles. There were spiny walkers with rows of wavy tentacles along their backs. These, however, represent branches of the tree that died out long ago, and we can only marvel at their fossils from afar.

In the Burgess fossils we see all the ingredients necessary for the tree of animal life's fantastic growth across the entire planet—and ultimately beyond—during the next 500 million years or so. We find strong skeletons against which complex muscles could operate to produce swift and precise movement. We find a host of different ways of obtaining food of all sorts, and a corresponding diversity of digestive equipment with which such food might be assimilated. We find sexual reproduction, with its accelerating evolutionary in-

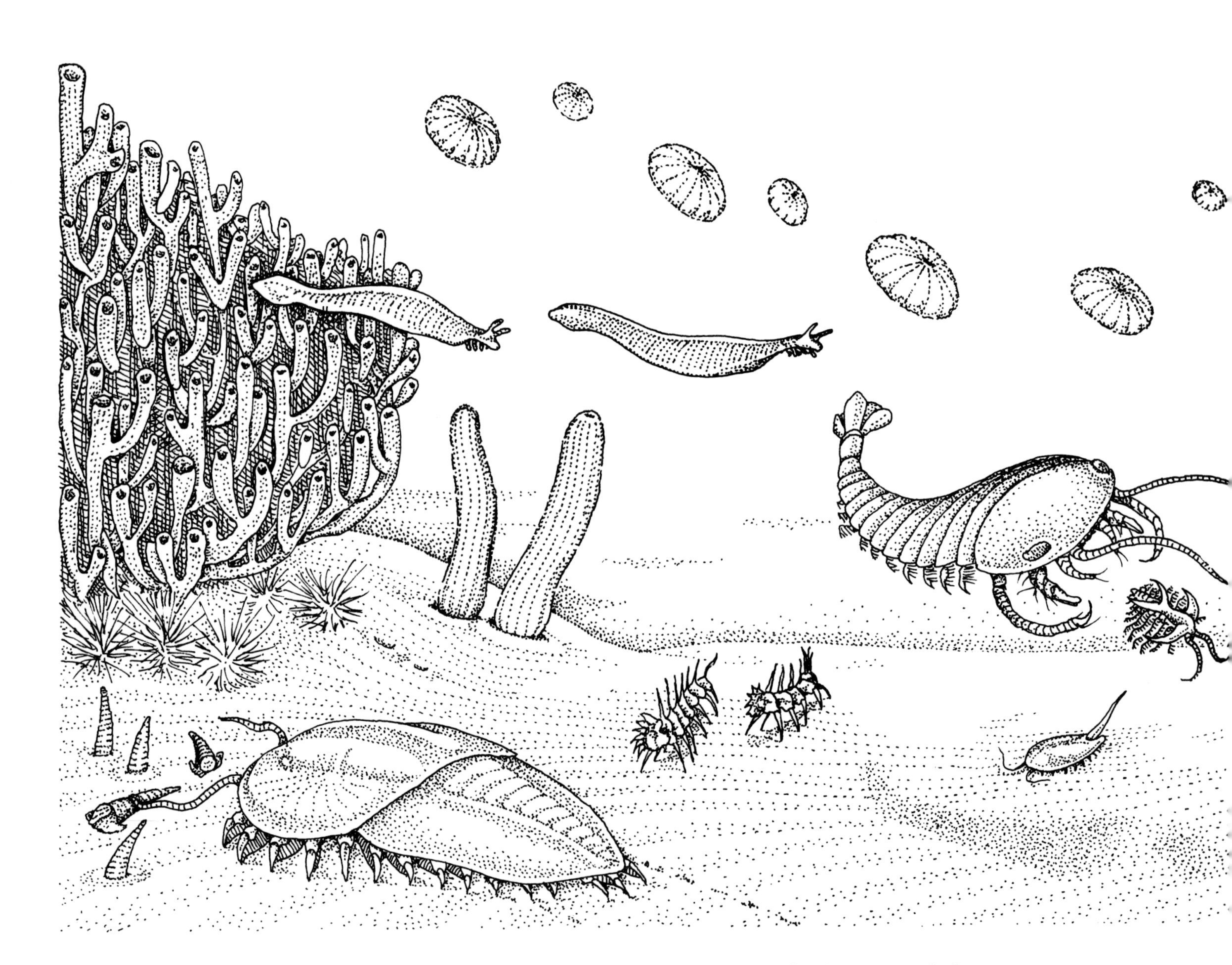

The Burgess sea bottom around 530 million years ago, showing some of the many animals experimenting with new ways of life. A trilobite (bottom left) *devours some burrowing mollusks. The elongated swimmers* (upper left) *are early chordates, close to our own ancestry in the tree of animal life. A large swimming arthropod attacks some*

fluence, almost universal among the metazoans. And all of these characteristics are, in at least some animals of the Burgess Shales, coordinated and directed by ever more sensitive and adaptable sensory and nervous systems. As a fauna (assemblage of animals as a whole), the Burgess fossils represent a point from which many boughs of our tree diverge to found modern ways of life and ecologic communities.

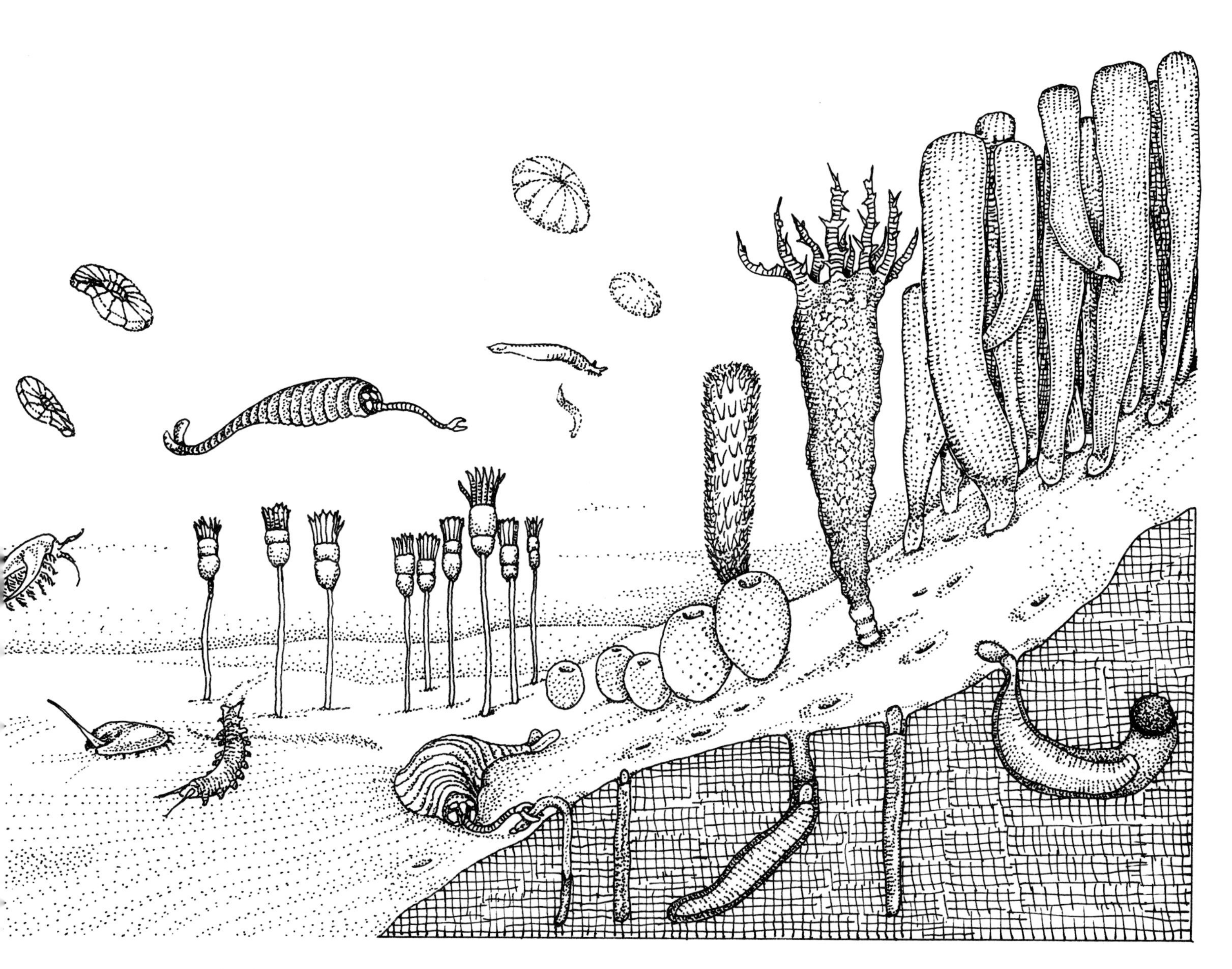

trilobites (left of center). *Below him walk mysterious forms with no known living relatives. Two strange five-eyed swimmers,* Opabainia, *are pictured, one swimming* (above) *and one pulling a roundworm from its burrow. The flowerlike beings rising from the sea floor are actually animals related to starfishes. Sponges frame the scene.*

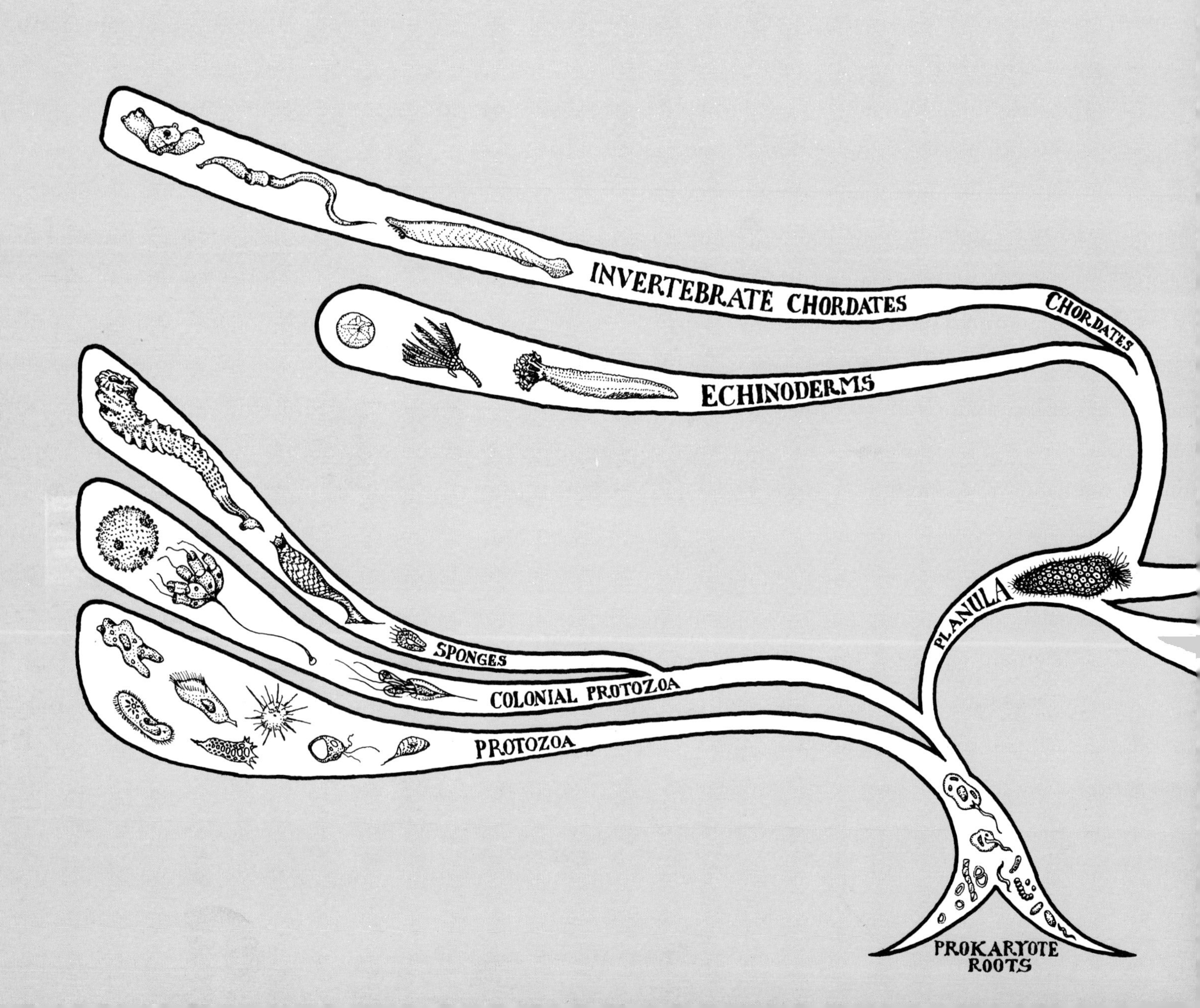
INVERTEBRATE CHORDATES
CHORDATES
ECHINODERMS
PLANULA
SPONGES
COLONIAL PROTOZOA
PROTOZOA
PROKARYOTE
ROOTS

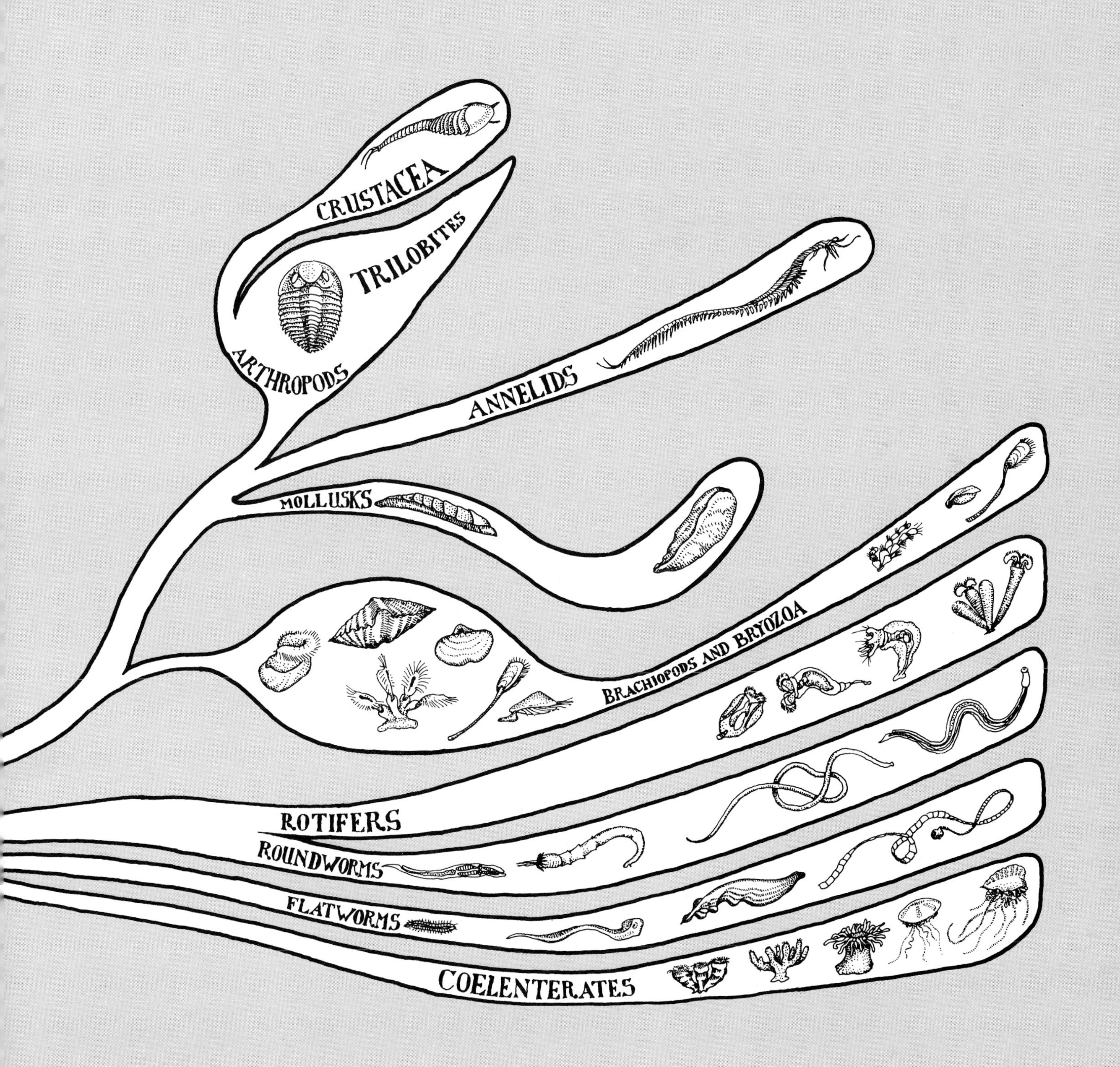
CRUSTACEA
TRILOBITES
ARTHROPODS
ANNELIDS
MOLLUSKS
BRACHIOPODS AND BRYOZOA
ROTIFERS
ROUNDWORMS
FLATWORMS
COELENTERATES

By around 425 million years ago, the fiercest creatures in the seas (and therefore on earth, since there was no land life as yet) were giant swimming arthropods that looked like scorpions. Often they reached three meters in length! These terrible animals, called eurypterids, or "wide wings," probably fed on an assortment of our own ancestors among the first vertebrates. The vertebrates were a group of chordates whose members had developed a series of little ring-shaped bones (called vertebrae) along their backs, which surrounded

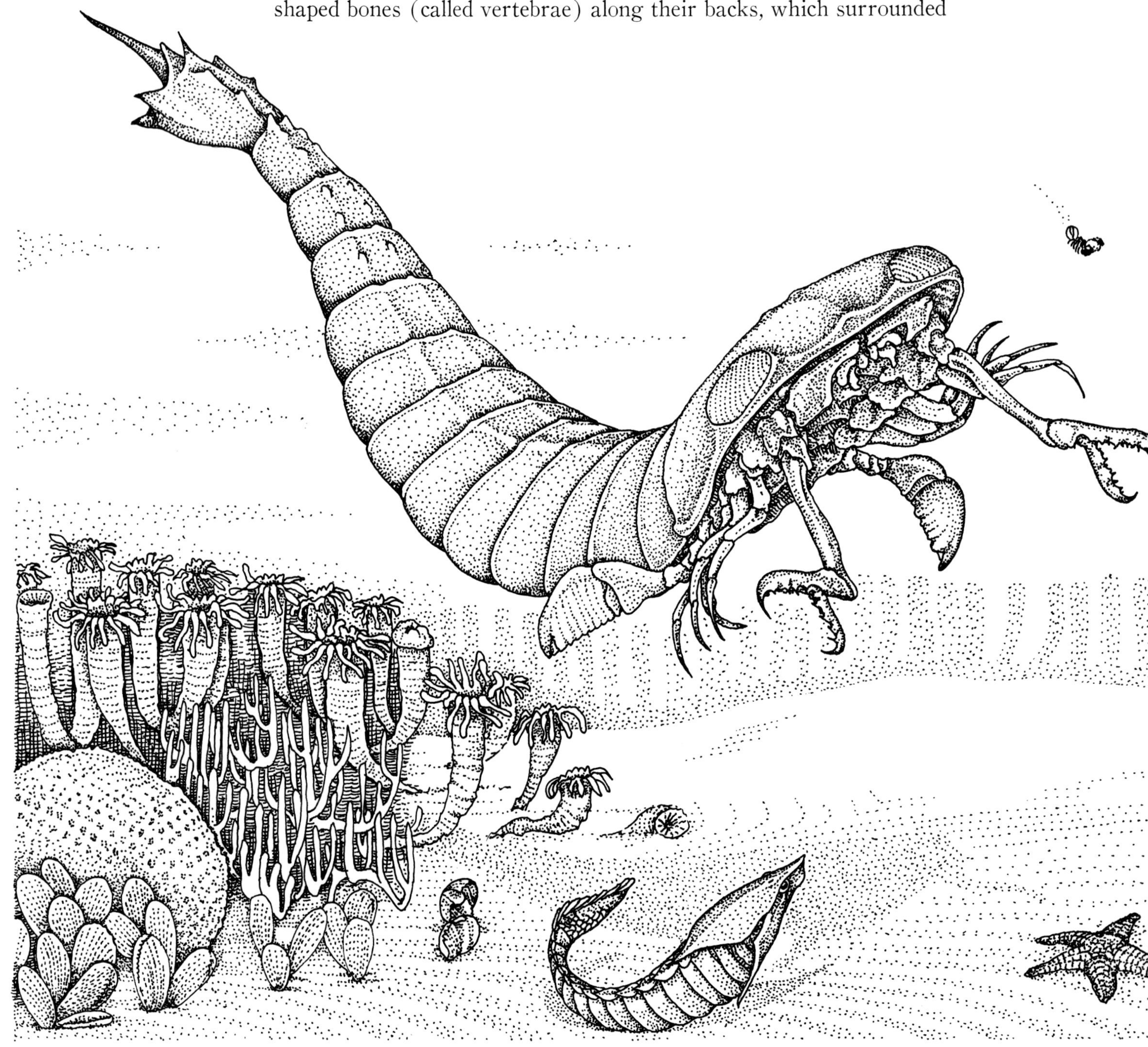

A sea-bottom scene from about 425 million years ago. On the left is a small coral reef with large individual coral polyps. Above it a eurypterid, a huge sea scorpion three meters long, prepares to scoop up an agnath, or armored jawless fish; such fish founded

and protected their all-important nerve cords. The early vertebrates were slow-moving, armored, fishlike beings with no jaws. Instead, their mouths acted as filters through which they strained small animals and other organic matter from the mud. They didn't look very impressive, but their descendants would do great things in times to come.

Other predators besides eurypterids hunted these lowly ancestors of ours. Nautiloids, mollusks distantly related to octopuses but encased in long pointed shells from which their heads and tentacles protruded, appear to have fed on vertebrates and trilobites and other slow-moving animals. But the nautiloids in turn were probably hunted by the mighty eurypterids, terrors of the seas.

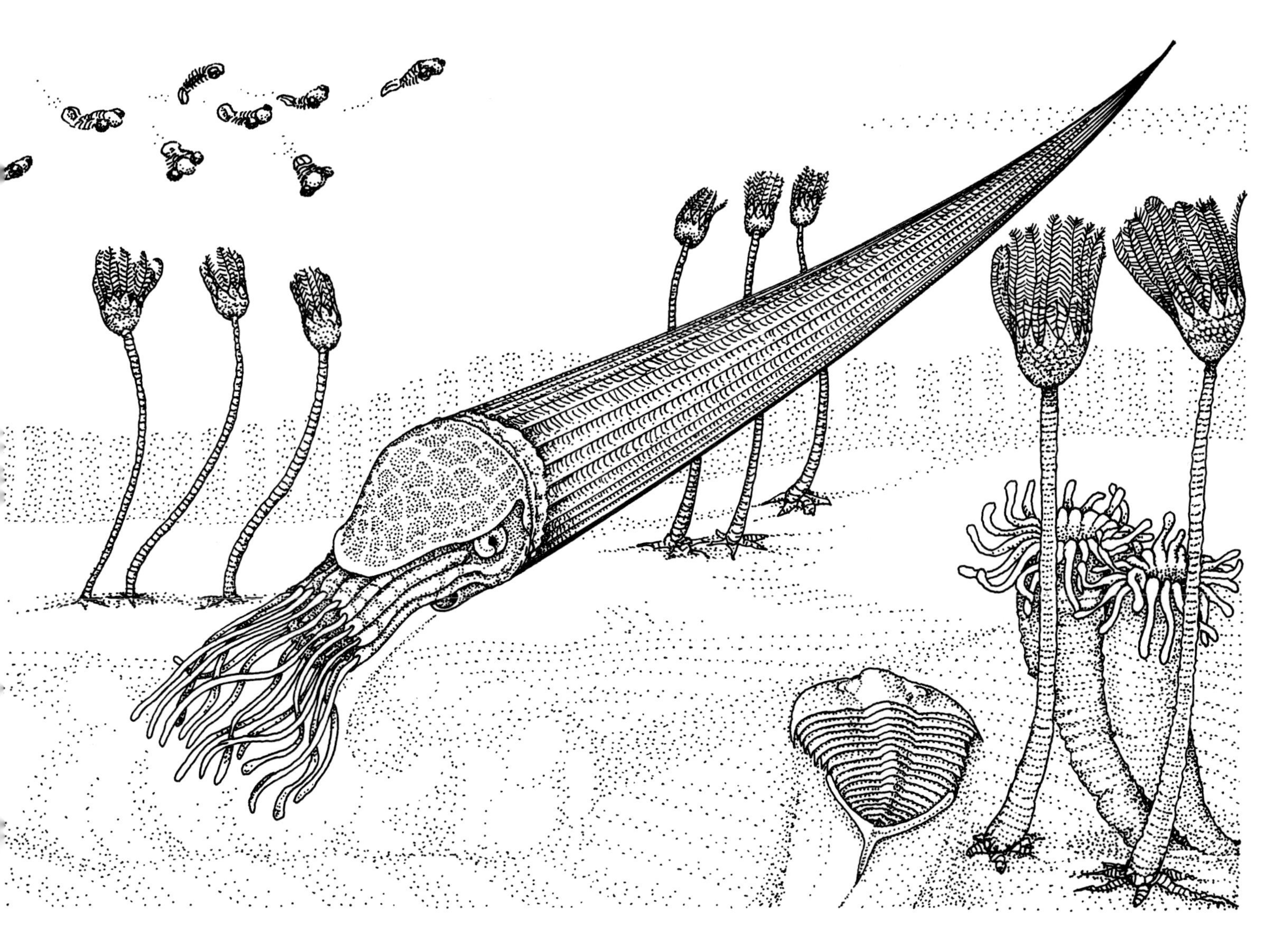

the vertebrate line. Tiny swimming trilobites flee the eurypterid's path, while below them a nautiloid, or shelled squidlike mollusk, swims away. The long-stemmed flowerlike creatures are crinoids. A trilobite grazes among crinoids and corals at lower right.

The rule of the eurypterids was doomed, however, by changes that arose among the primitive vertebrates they hunted. These vertebrates possessed gills through which they procured dissolved oxygen from the water and strained their food from the mud. Mutations among early vertebrates produced a few models whose forward gill arches became hinged, movable jaws. This was a change that permitted them to eat more solid food than their filter-feeding ancestors had been able to. A host of new econiches opened up for these jawed vertebrates, and they assumed new roles toward their neighbors. Many became fierce predators with muscular, streamlined bodies that allowed them to swoop down upon and eat even the mighty eurypterids.

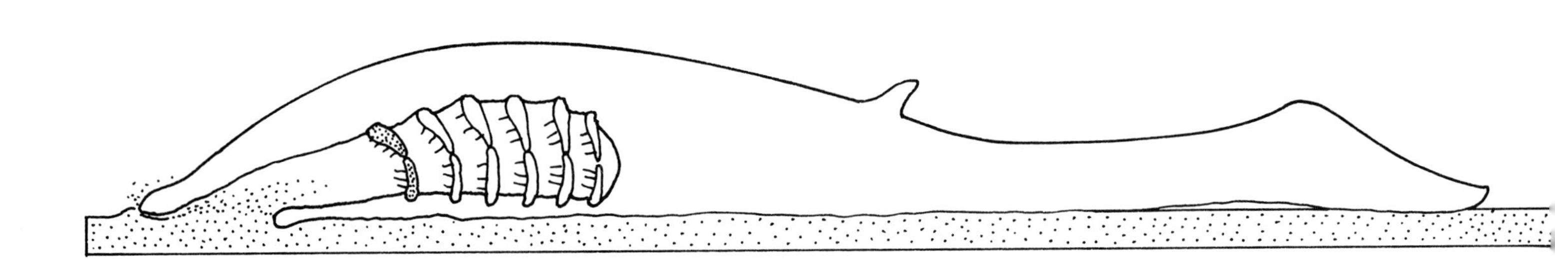

The evolution of jaws. (above) *A bottom-dwelling agnath, a primitive vertebrate which lacked jaws, filters its food from the mud. Its forward gill arches are shaded.* (below) *A primitive jawed fish. Its jaws, shaded, show little differentiation from gill arches, but are nonetheless highly functional, enabling the fish to engage in active predation.*

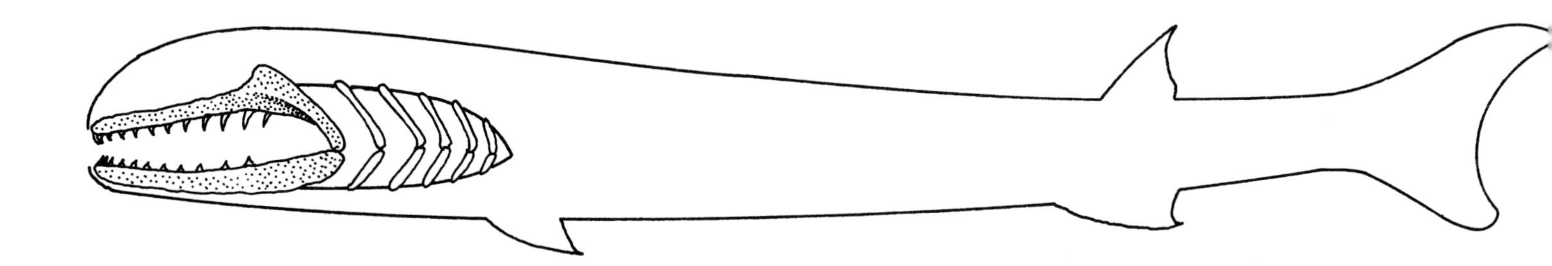

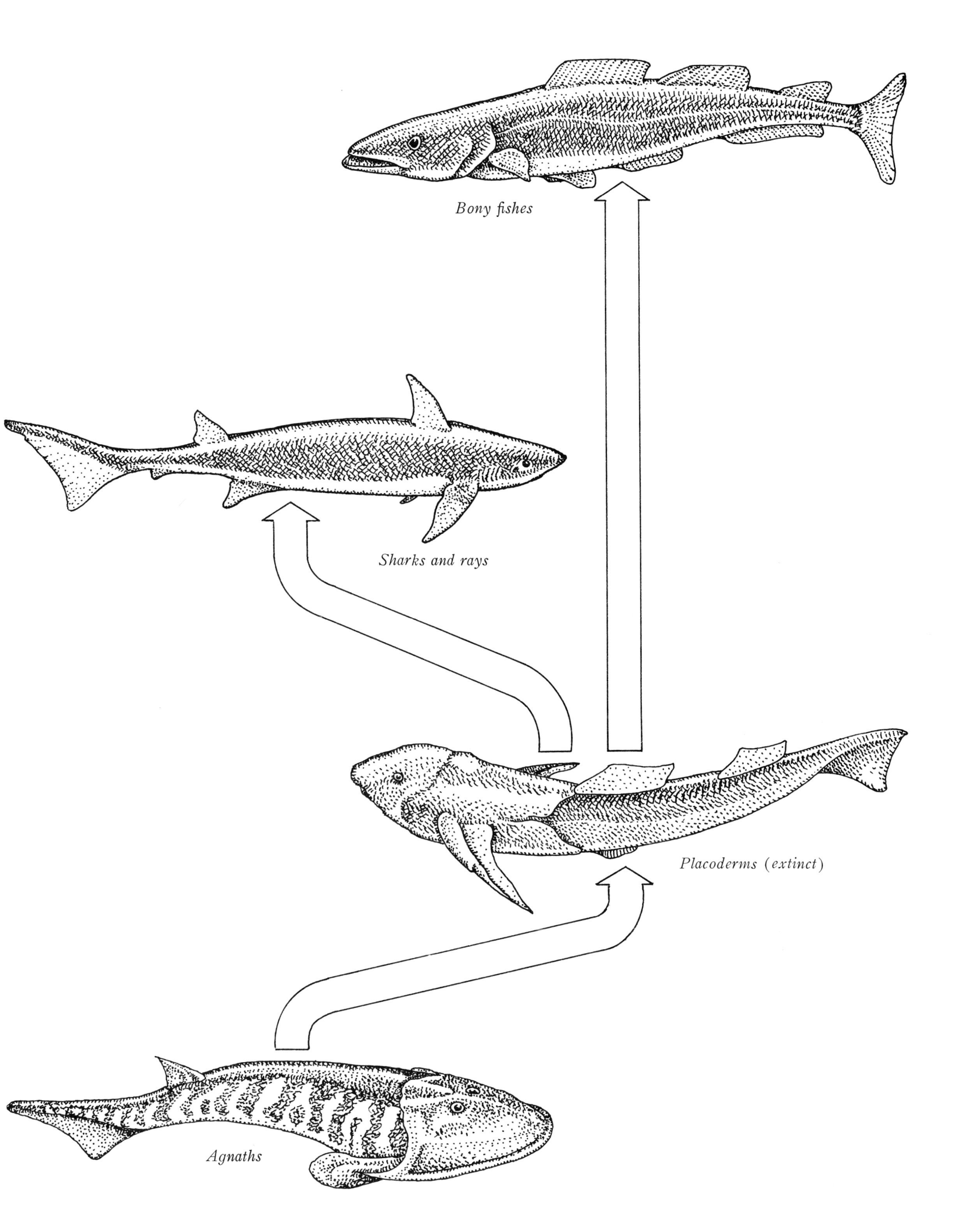

THE ADAPTIVE RADIATION OF FISHES FROM THE ORIGINAL JAWLESS FORMS.

Around 415 million years ago, these new jawed vertebrates, the first real fishes, were replacing the giant scorpions as kings of the ocean. Some of these early fishes grew to ten meters in length, perhaps too big to swim very fast. Such giants were heavily armored and may have lain about like great steel traps until some unfortunate smaller animal swam within reach of their fearsome jaws. The ancestors of sharks appeared during this interval, the first fish with skeletons of cartilage rather than bone. Sharks and their relatives are conservative—they have changed little in 400 million years—but they are also beautifully efficient, the reason they have been so successful through such a vast span of time.

The shallow sea bottom of around 413 million years ago. (left to right) *An ammonoid, a fast-swimming mollusk related to the modern chambered nautilus, floats next to a*

Although trilobites continued to outnumber all other forms of life on the sea bottom, the jawed vertebrates and other predators caused these slow crawlers a great deal of trouble. New mollusk forms, including fast-swimming ammonoids (nautiloid-like tentacled predators with coiled shells), competed with vertebrates and with one another for the rich trilobite food. Trilobite populations began to diminish.

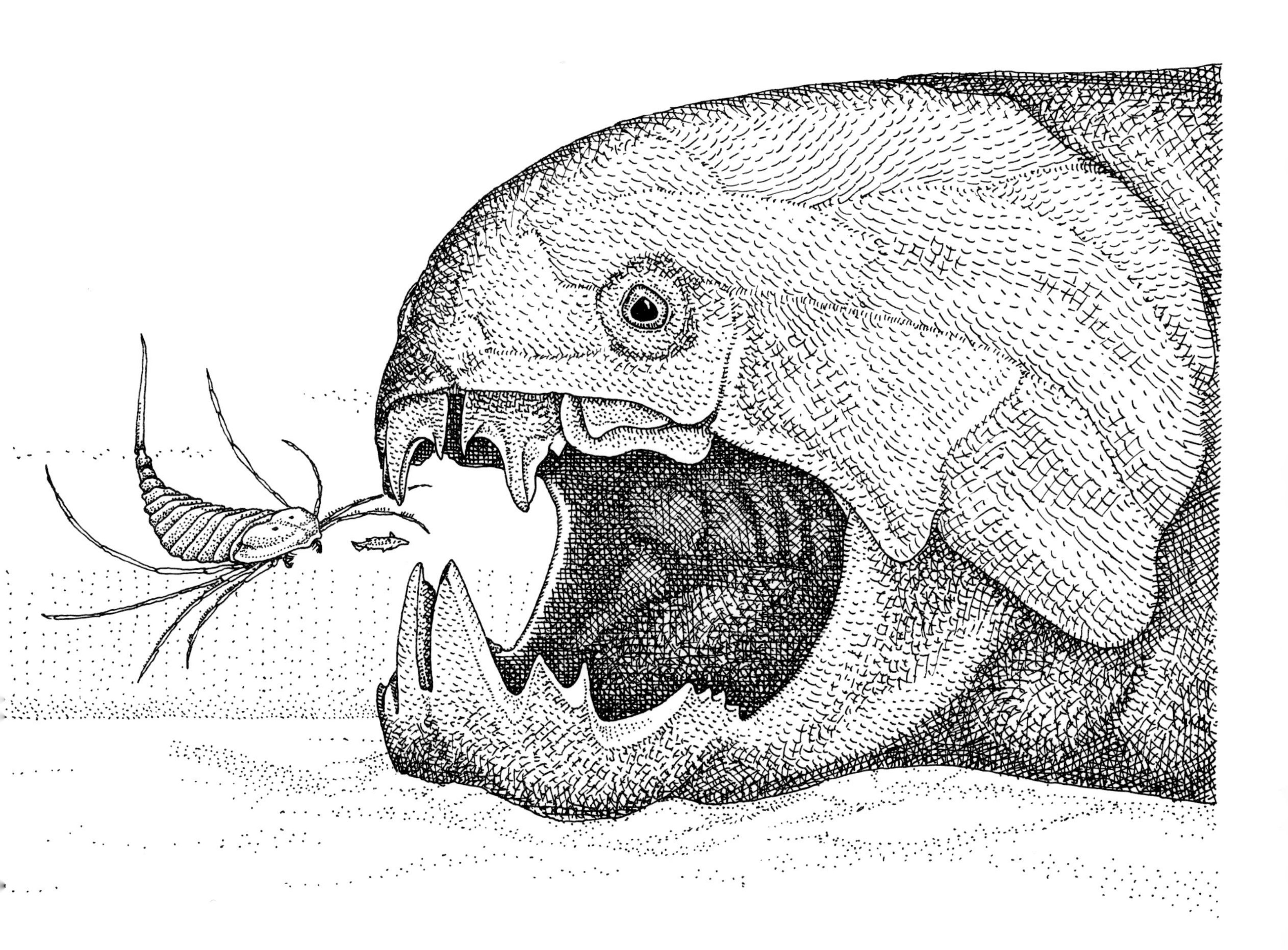

pair of small, primitive, jawed fishes. (right) *A gigantic armoured fish prepares to catch a eurypterid, itself chasing a smaller fish. The rise of jawed vertebrates signaled the end of the sea scorpion line.*

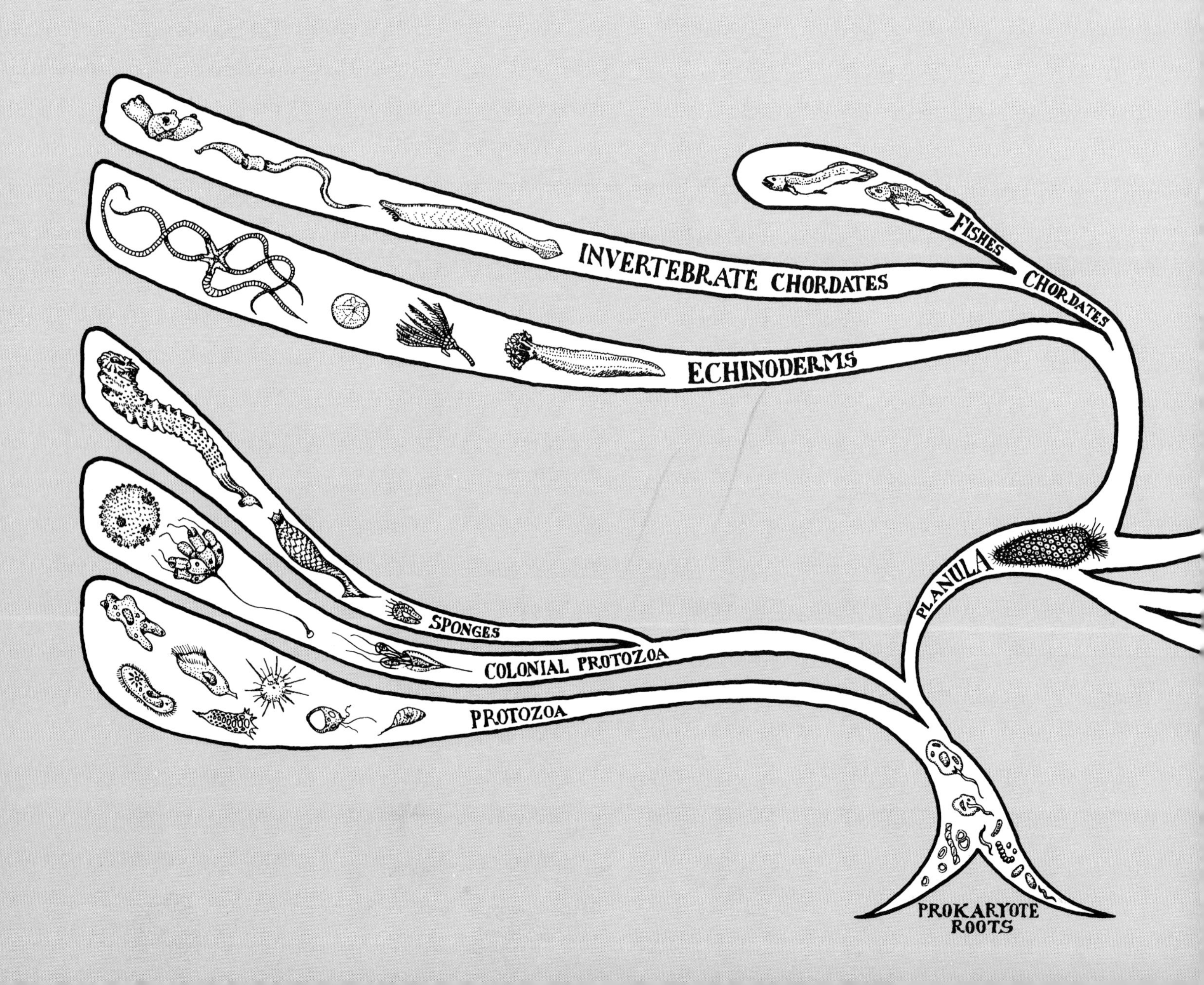

INVERTEBRATE CHORDATES
FISHES
CHORDATES
ECHINODERMS
PLANULA
SPONGES
COLONIAL PROTOZOA
PROTOZOA
PROKARYOTE ROOTS

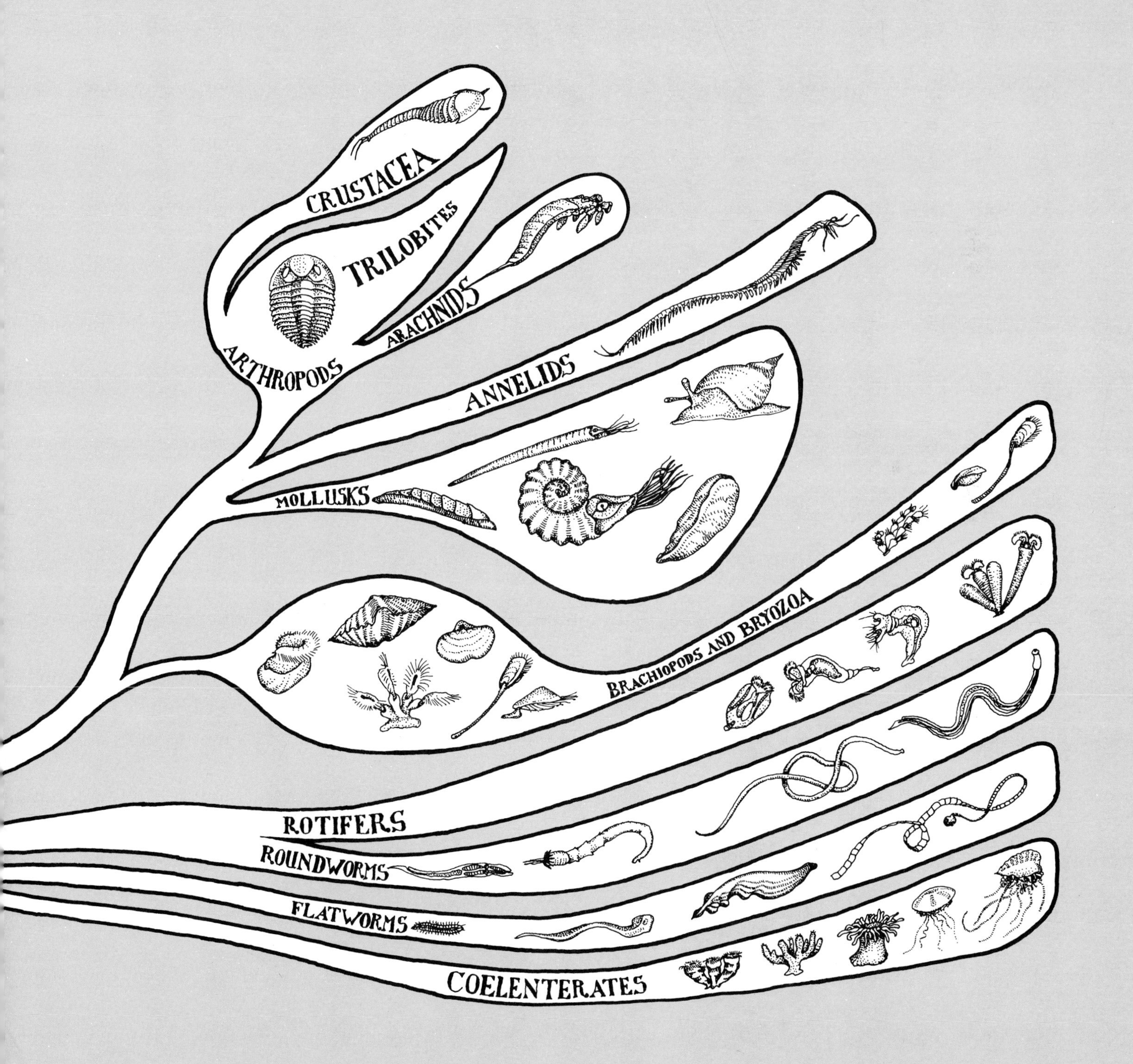
CRUSTACEA
TRILOBITES
ARACHNIDS
ARTHROPODS
ANNELIDS
MOLLUSKS
BRACHIOPODS AND BRYOZOA
ROTIFERS
ROUNDWORMS
FLATWORMS
COELENTERATES

Trilobite descendants, scorpion-like and crablike animals, were being squeezed to the edge of the water by the teaming competition in the seas and in freshwater. Competition for survival was at its most intense at the shorelines of the world. Here, abundant sunlight and minerals washed from continental rivers created a uniquely nutrient-rich environment in which many life forms flourished. It was here that one of the most important revolutions occurred. Life overcame the shoreline barrier and invaded the land.

Along the water's edge a new world of plants had become established. These were the pioneers of life on land, an evolutionary leap so great that we might compare it with our own relocation to another planet. Life in the deep was always surrounded by the water in which all metabolism takes place. Underwater there is no danger of drying up. On land, however, things *do* dry up—water evaporates into the surrounding air—and metabolism ceases.

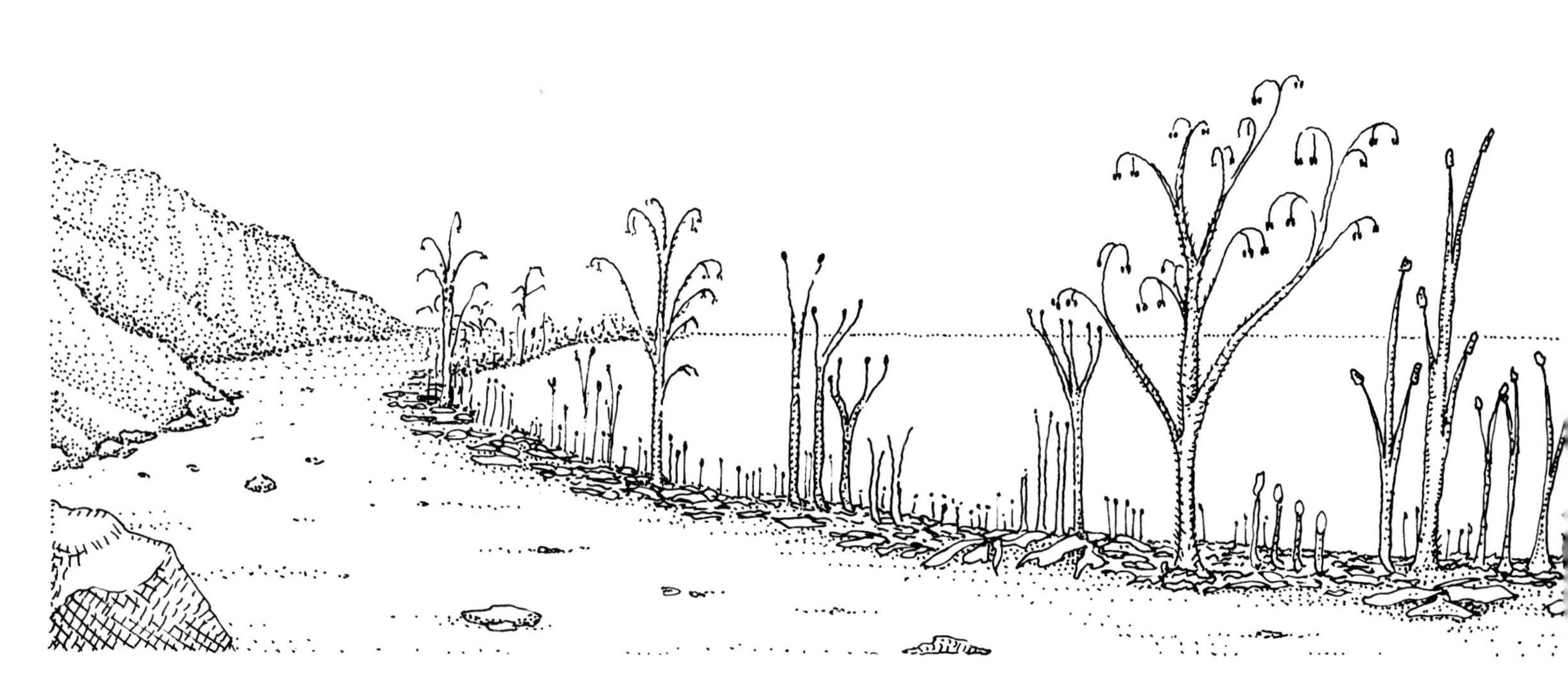

Plant life invades the formerly lifeless land around 415 million years ago. Early pioneer plants still had to be near water in order to reproduce. In this shore scene, ancestors and relatives of modern liverworts, whisk ferns, and mosses compete for space along the edge of the land. Such humble plants were the food supply for early land animals.

In fact, because of the great problems associated with life in air, members of only five of the many phyla of living beings have successfully made the transition. These are the Fungi (mushrooms and molds), the Tracheophyta (vascular or higher plants), certain mollusks, arthropods, and vertebrates.

Because all living animals feed on solar energy trapped by green plants (either by eating plants directly or by eating animals that have eaten the plants), it is logical to suppose that green plants were among the first living things to colonize the land surface. They appear to have done so in the form of bryophytes, mosslike plants that can live on land but depend on the presence of water for their reproductive processes. This is why mosses and their relatives tend even today to grow in moist places.

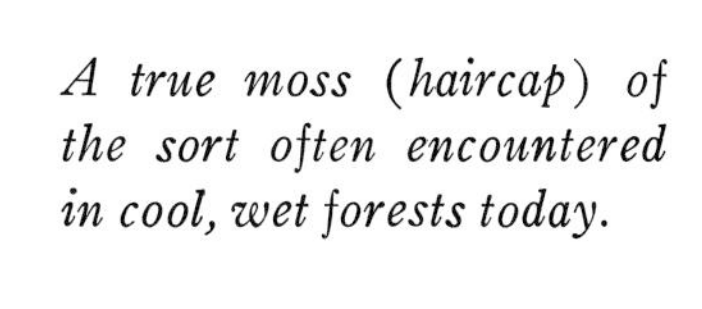

A true moss (haircap) of the sort often encountered in cool, wet forests today.

MODERN FORMS OF EARLY LAND-DWELLING PLANTS.

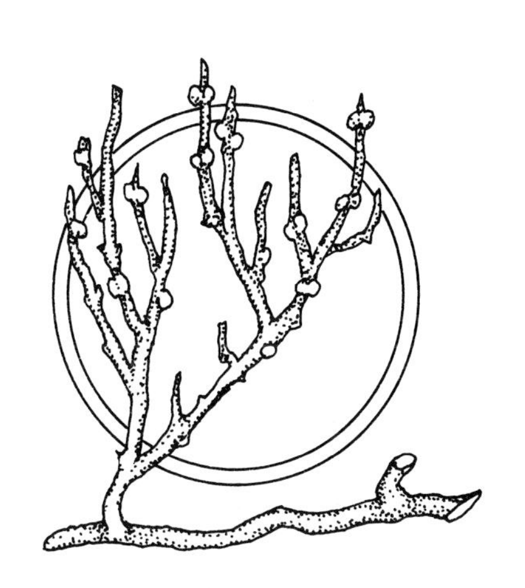

A whisk fern, Psilotum, *inhabitant of moist lowlands in the tropics.*

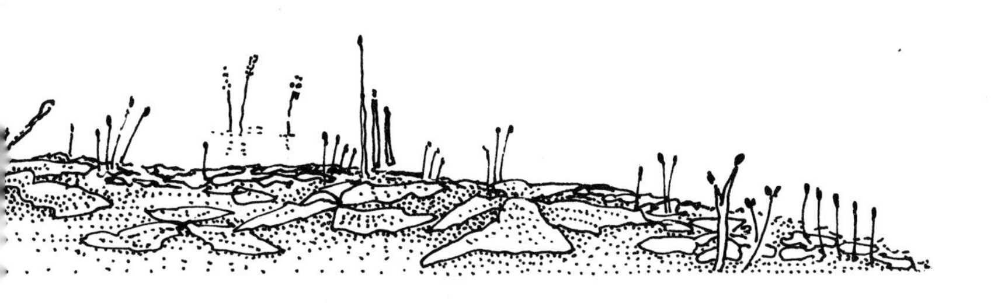

A liverwort, Marchantia, *found along moist streambeds today.*

In all likelihood, the world of 415 or so million years ago was bright green along the edges of most bodies of water as the bryophytes prospered. The selective pressures the plants experienced during periods of dry weather, and from evaporation, forced them to evolve, and they developed waterproof cuticles that protected their own internal water. These cuticles also afforded them some resistance to the thousands of herbivorous animals that shared the shallow waters with them. Meeting no competition on land, generation after generation of these pioneering bryophytes radiated onto marine and freshwater shores, and at some point produced the first vascular plants. Vascular plants have real tubes (veins) through which they protect and transport their vital moisture and nutrients.

These vein-bearing plants were completely unimpeded by any vegetarian animals on land. After all, no animals were there yet. So, by about 410 million years ago, much of the earth's land surface was covered by mighty forests of ferns, trees, and other vascular plants.

As we have seen, plants are the only forms of life on earth capable

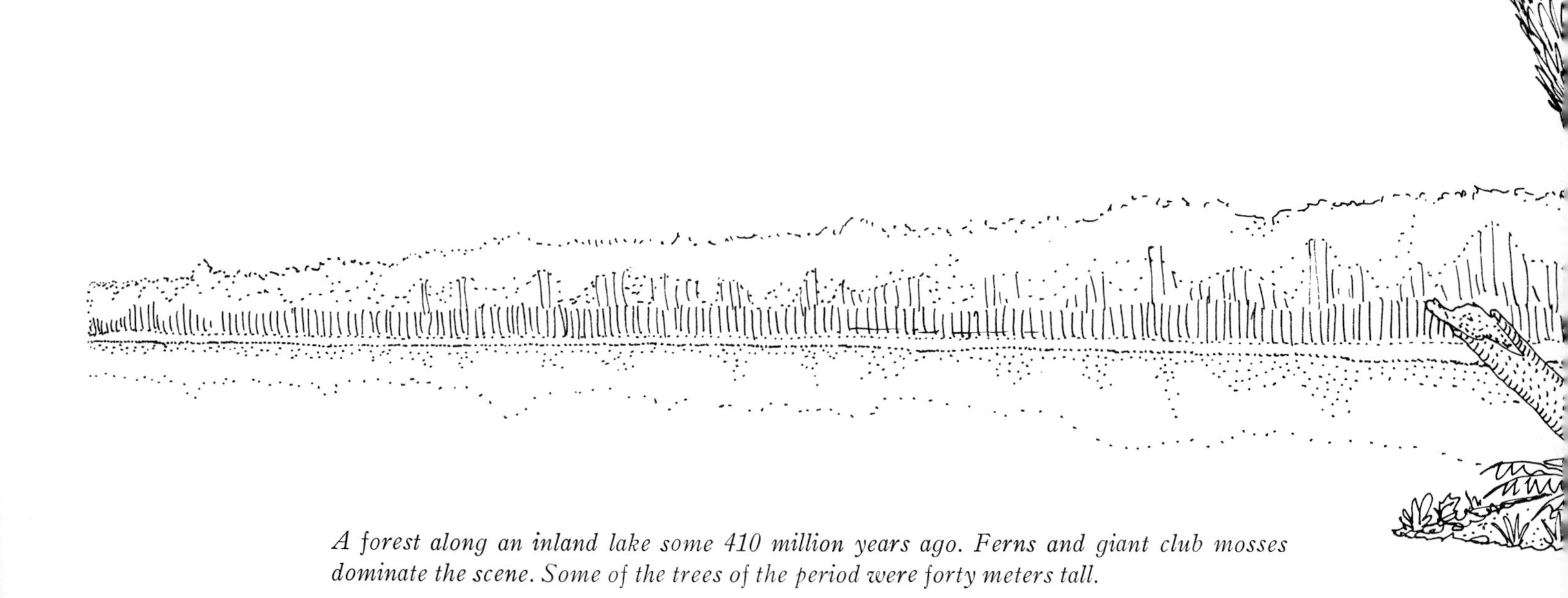

A forest along an inland lake some 410 million years ago. Ferns and giant club mosses dominate the scene. Some of the trees of the period were forty meters tall.

of producing living matter, or biomass, from nonliving matter (minerals) and the sun's energy. The immense forests of 410 million years ago represented billions of tons of biomass that would more than nourish any animals that might exploit it. Given all the competition, predation, and other selective pressures operating on the teeming animal life in the deep back then, it is hardly a wonder that certain of the more hard-pressed animals did not take refuge in ever more shallow water.

The first animal colonizers of land for which we have fossils were small arthropods. They seem to have looked very much like modern-day scorpions. Their water-dwelling ancestors, eurypterids and their kin, already possessed features that could prove useful on land, including a rigid external skeleton. With only minor modifications, that exoskeleton could both protect them against drying out and support them in the absence of the buoying help of water. And, most aquatic arthropods had (and still have) legs, which enabled them to get about without having to float, as do, say, fishes.

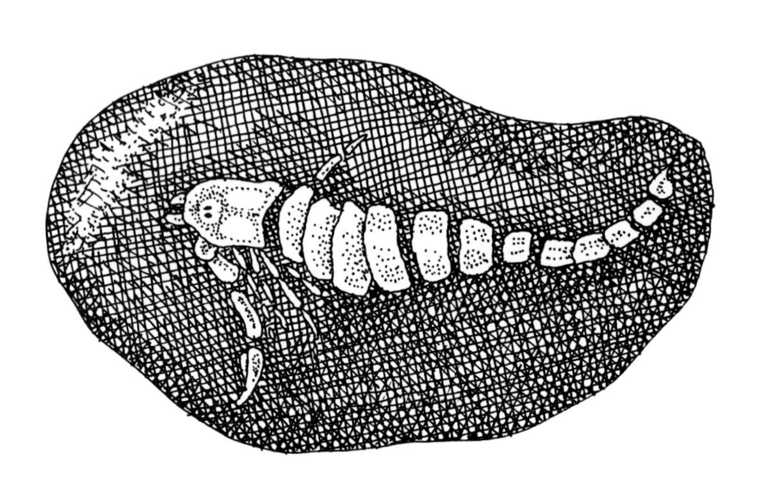

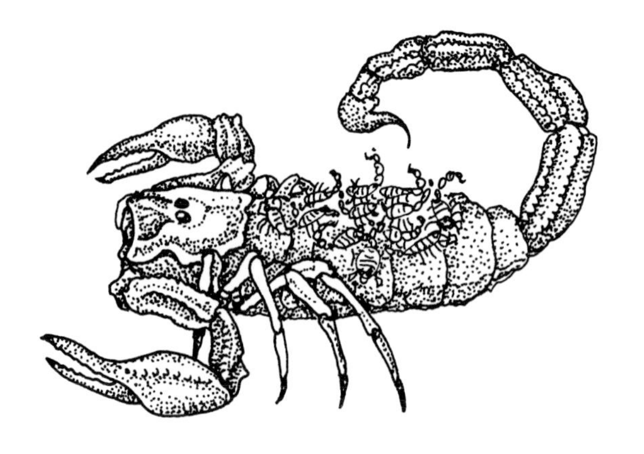

Fossil scorpion (left) *and living scorpion. Scorpions have not changed much in four hundred million years.*

Fossils of scorpions and their relatives in the great arthropod class Arachnida date from 415 million years ago. The scorpions have changed so little in that gigantic span of time that we can say these animals are very, very good at living their conservative sort of lives. Equipped with a potent sting to stun their prey or discourage their predators, scorpions are nearly immune to unwanted attention from most other animals (except human beings, who always try to squash them). In addition, they take good care of their young, protecting their eggs and carrying their babies on their backs. And they are adept at sneaking about and remaining almost invisible. They are, in fact, among the most successful animals on our planet and may thus be considered things of beauty and ecologic wisdom. It is a pity that so many people hate these wonderful creatures.

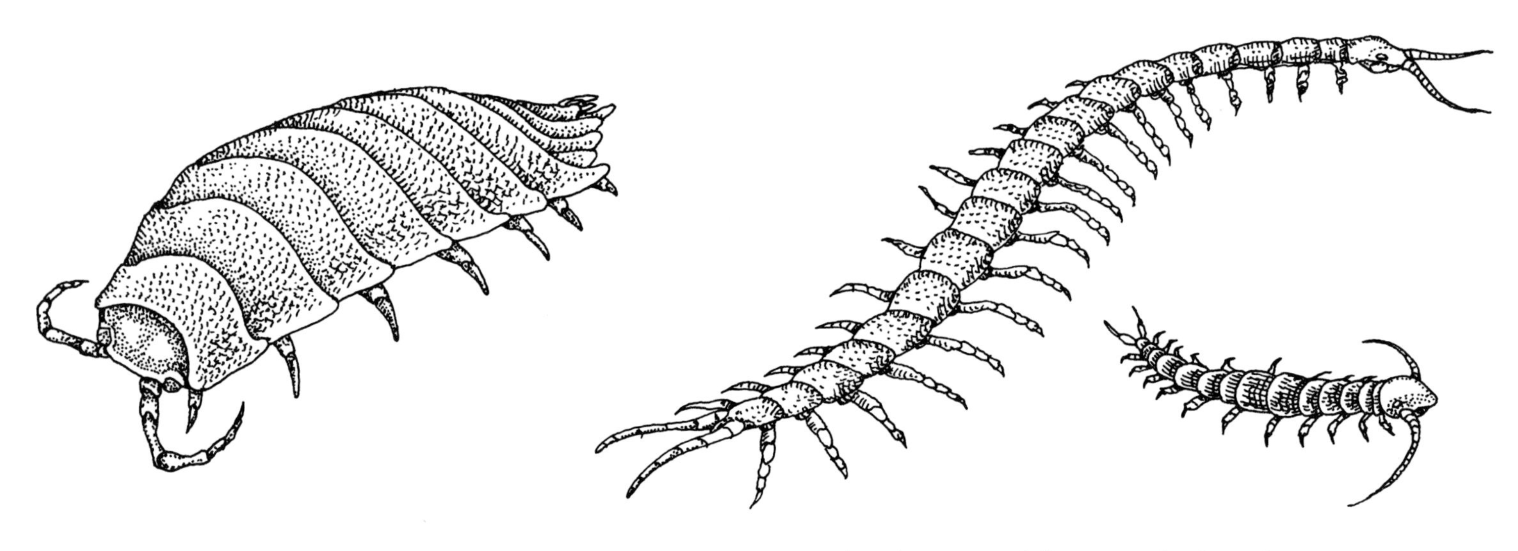

Among the first arthropods to adapt to land-dwelling scccessfully were the isopods (left), *or sowbugs. These may be found under rocks today, little changed through hundred of millions of years of evolution.* (middle and right) *A centipede and a symphylan (a small garden centipede). From a symphylan-like creature may have evolved the first insects.*

From those early scorpions eventually descended mites, spiders, and other members of the class Arachnida. Most of these were predators, and their fossils imply that by the time they evolved, some herbivorous animals must have already arrived on land, on which the arachnids fed. This was indeed the case. Crustaceans (relatives of shrimps and crabs) like modern pillbugs and sowbugs had also come ashore as herbivores and scavengers (eaters of dead organic matter). From animals like these may have descended the vegetarian millipedes and their predatory relatives, the fast-running centipedes. These myriapoda (many-legged creatures) made ever longer forays into the great forests of the time.

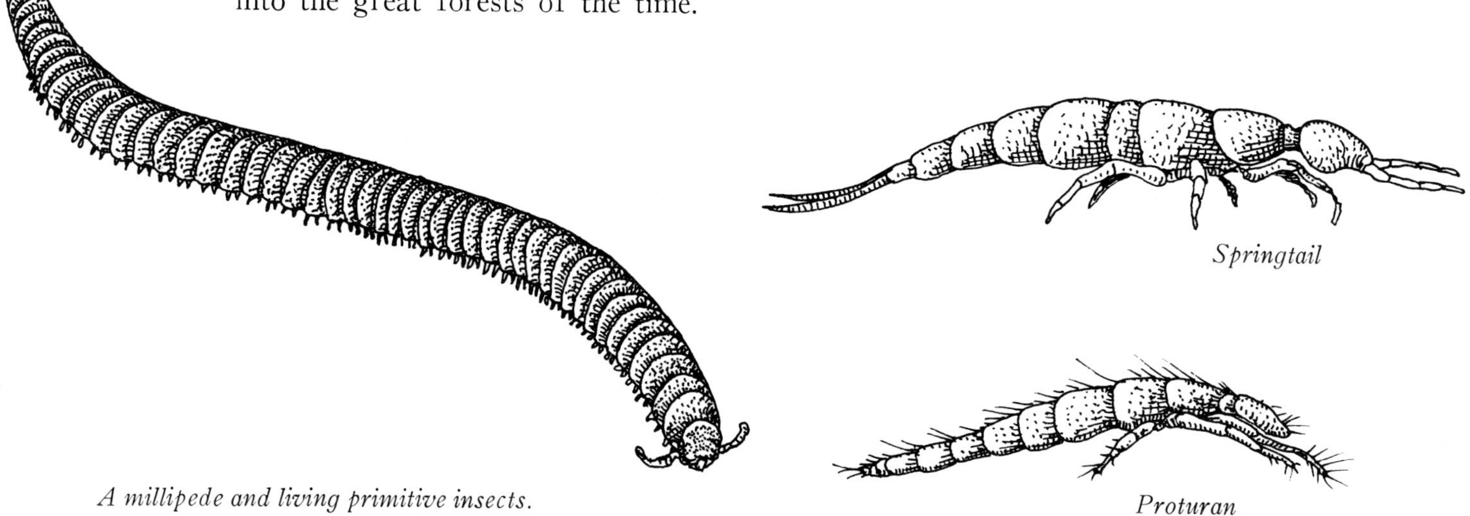

Springtail

Proturan

A millipede and living primitive insects.

From millipedes or centipedes the first insects arose, probably about 413 million years ago. They were early representatives of the class that would ultimately dominate the world of land animals. The first insects, like their ancestors, were fast ground-crawlers. But unlike the centipedes and millipedes, which have many legs and many body segments, the new insects possessed only three pairs of legs and a simplified body plan that consisted of three parts: a head, a thorax (on which the legs were fixed), and an abdomen containing the organs of digestion and reproduction.

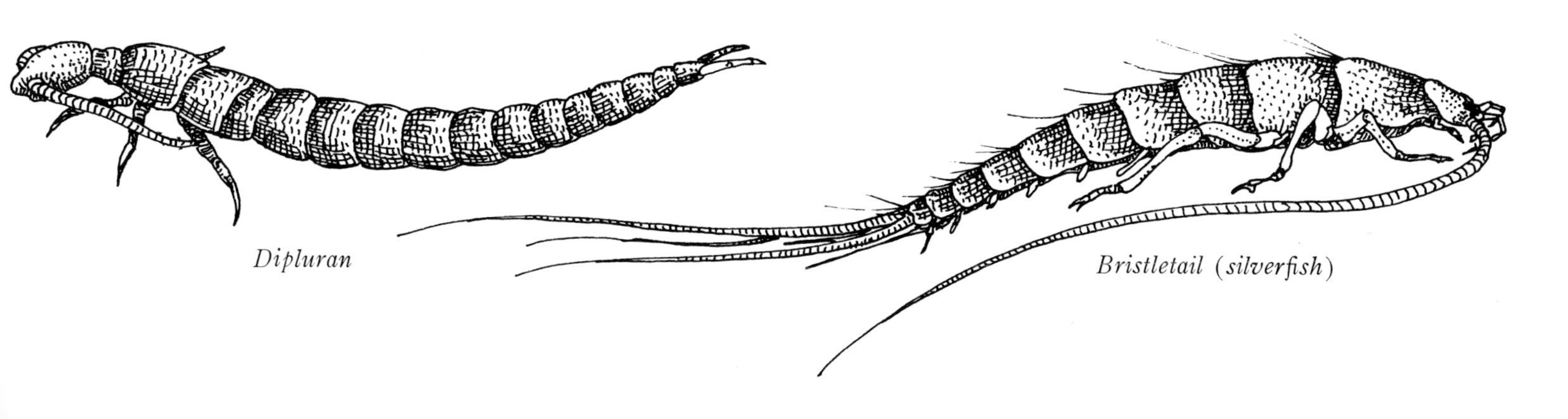

Dipluran

Bristletail (silverfish)

Because insects are small, they have always been eaten, squashed, and otherwise picked on by larger animals. But because they produce large numbers of young, laying as many as several hundred eggs at a time, they are usually able to circumvent these pressures. Insects evolve far faster than do most land animals. The frequency of their egg-laying and the inevitable small differences among their many young permit some to invade new econiches faster than others. Any two insects that survive a particular selective pressure may leave *hundreds* of offspring which inherit the characteristic that enabled their parents to survive.

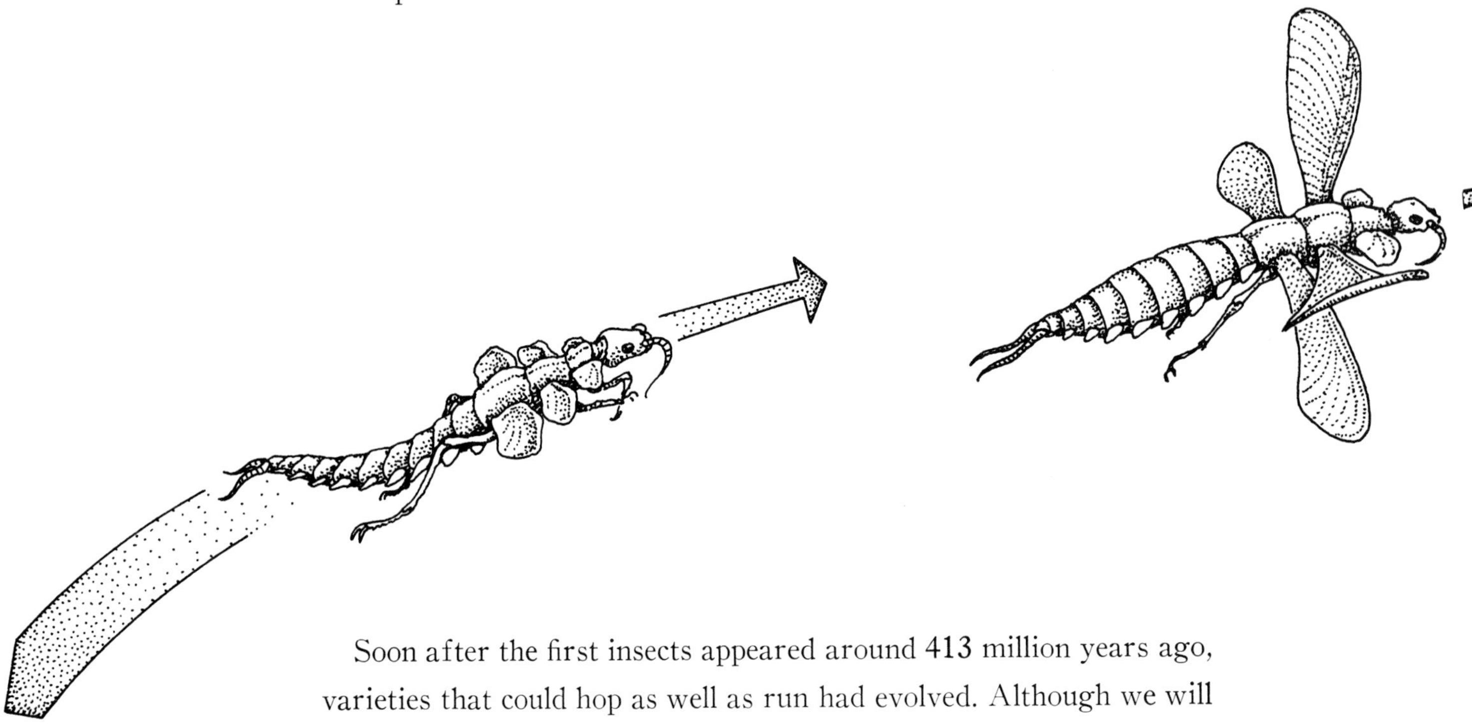

Soon after the first insects appeared around 413 million years ago, varieties that could hop as well as run had evolved. Although we will never know precisely when, some mutants appeared possessing flaps of chitin (the skeletal material of insects) on their thoraxes above their legs. While these flaps may originally have had some other purpose, they seem eventually to have served as gliding planes, extending the hops and bounds of the insects that possessed them.

Because the better gliders tended to escape their predators, the ability to glide became gradually enhanced until true flight appeared. Rather quickly, on the slow evolutionary time scale, insects that could fly were able to spread across the land and take up life in places unavailable to any other animal life on earth. Insects live in lichens, in hot springs, and are usually the first animals to colonize lands blasted by volcanoes. They live as parasites all over plants and even on one another. The world became, and has remained so ever since, a singing, buzzing nation of insect life.

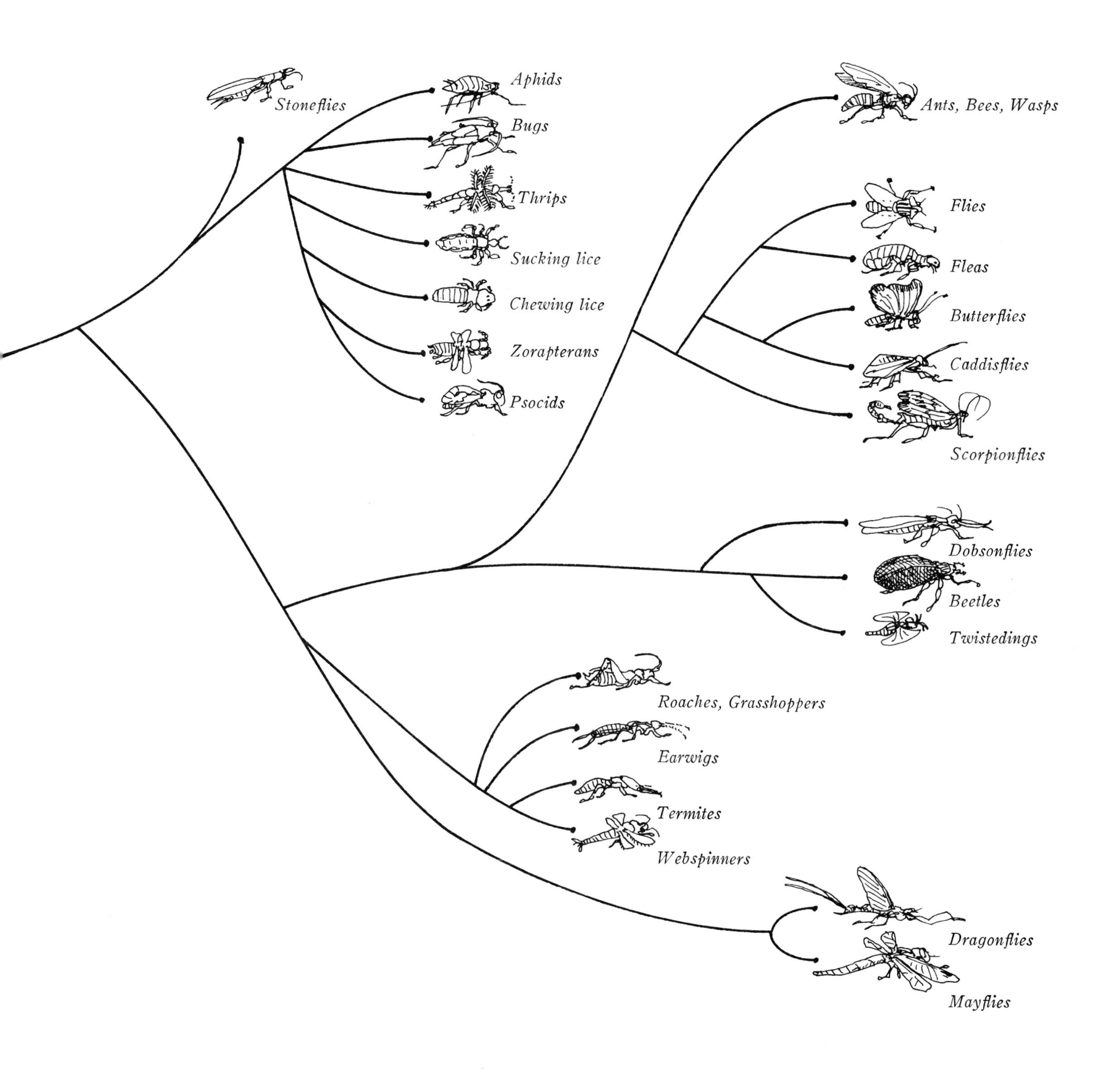

Stages in the evolution of insect flight. Once insects evolved the capacity to fly, they experienced an explosive adaptive radiation that continues to this day. (opposite page, left) *A primitive (hypothetical) leaping insect whose thorax, or chest region, supports vanes for heating and cooling. It gives rise to an early true flyer, whose fossils show a vestigial remnant of the third thoracic pair of vanes. From some such early aeronaut evolved all of the many orders of flying insects.*

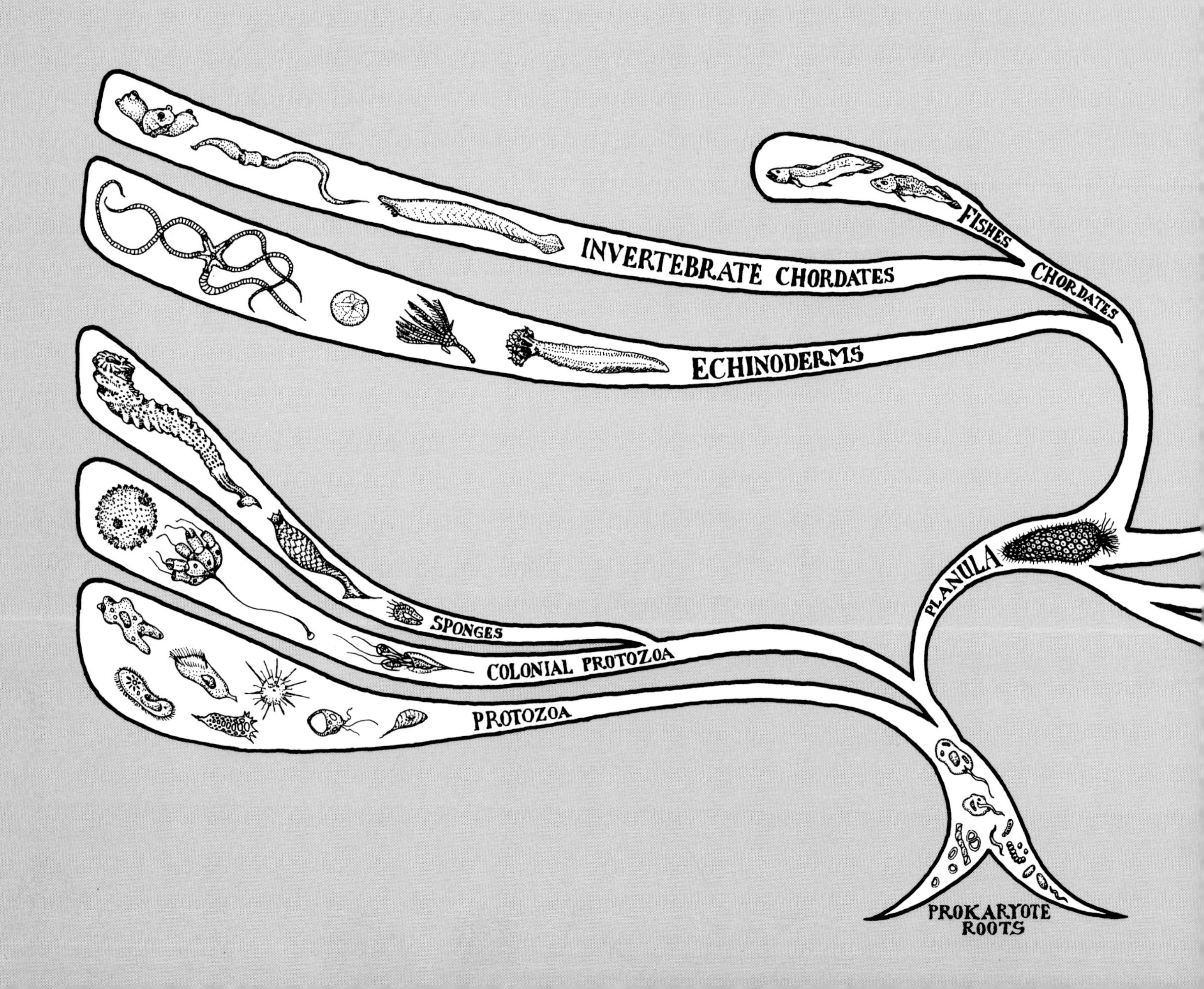

INVERTEBRATE CHORDATES
FISHES
CHORDATES
ECHINODERMS
PLANULA
SPONGES
COLONIAL PROTOZOA
PROTOZOA
PROKARYOTE ROOTS

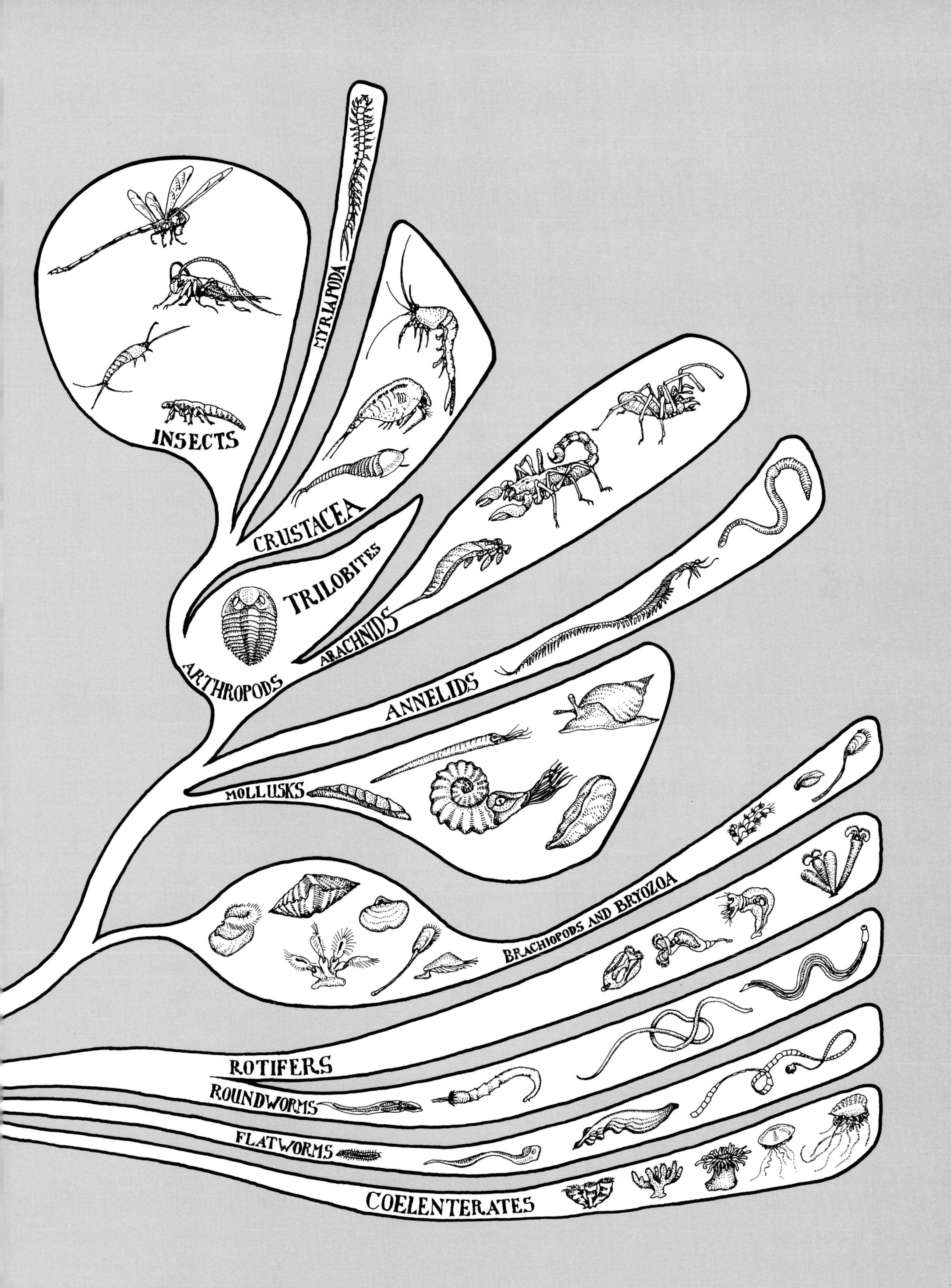

INSECTS
MYRIAPODA
CRUSTACEA
TRILOBITES
ARACHNIDS
ARTHROPODS
ANNELIDS
MOLLUSKS
BRACHIOPODS AND BRYOZOA
ROTIFERS
ROUNDWORMS
FLATWORMS
COELENTERATES

Meanwhile, back in the water, the fishy vertebrates were filling up all the econiches available to them. Moving from open rivers and lakes into shallow swamps, various fishes occasionally became exposed to drying, as the water levels of these swamps fluctuated. Most such fishes died, being, after all, creatures of water. Nevertheless, a series of lucky mutations somewhere produced fishes with a lung or two. These lungs were internal sacs in which oxygen from the air might be moistened with body water and passed into their bloodstreams.

Fishes possessing such lungs could survive the drying of the swamps for a time, but they were not able to live in complete dryness for long—they were still water-dependent. Some means of adapting to constantly fluctuating swamp levels was necessary if they were going to survive—and some fishes were lucky enough to be endowed with a means.

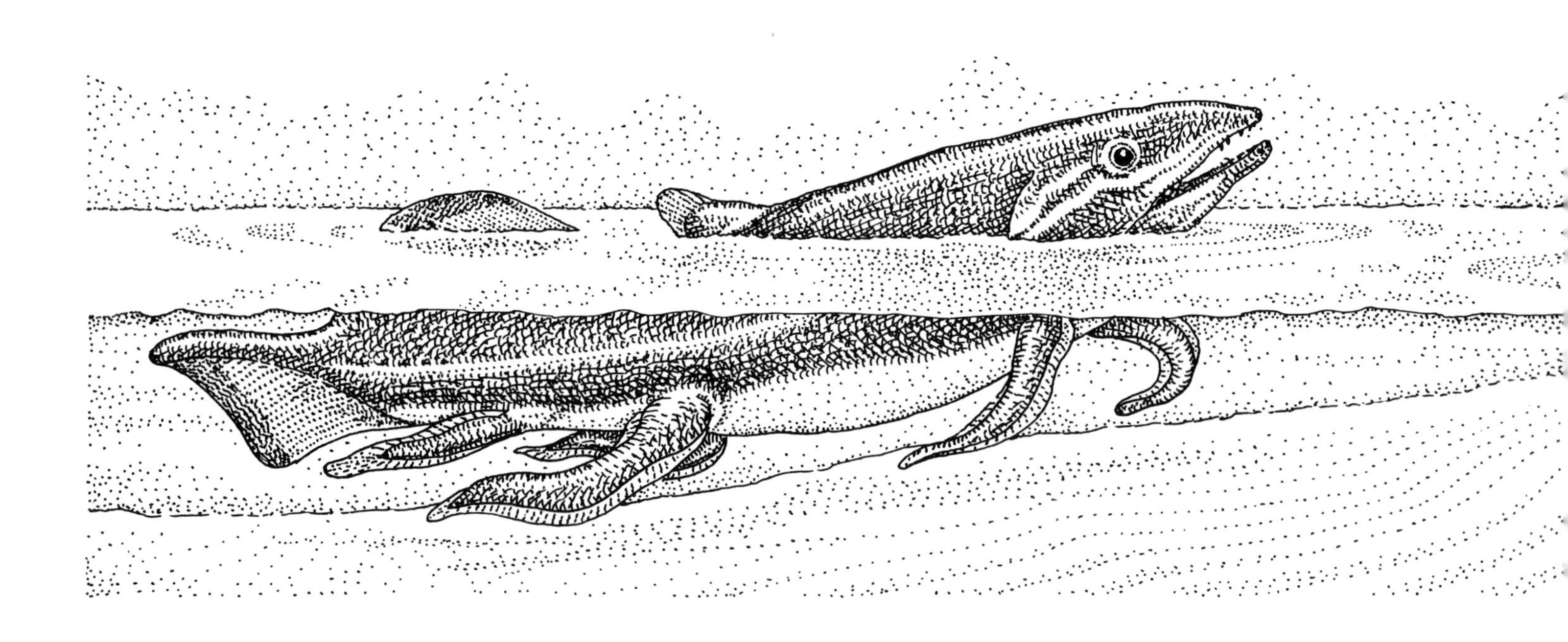

Early lungfishes with muscular fins were able to seek better water when their home ponds dried up or became too stagnant for survival. It is important to realize that such fishes

Fishes with lungs also possessed fins (present in the earliest invertebrate chordates), which moved them about in water. Some had more muscular fins than others. In the struggle to find more and better water during dry spells, the fishes with strong fins were able to push themselves through the mud to a swamp that contained enough moisture to sustain them. Those with more muscular fins had greater success at this than those with weaker fins. These muscular-finned survivors mated with each other, producing offspring with fins equally suitable for pushing. By about 410 million years ago we find real crawling vertebrates moving about in wetlands, descendants of fishes whose fins had evolved into stubby legs to take them from swamp to swamp.

Such crawlers were still fishlike in habits, but they had one great advantage: the option of traveling on land to find better swamps in times of drought. This advantage permitted them to live in a variety of shallow bodies of water. Because they were essentially of two worlds, water and (to a lesser extent) land, we call these beings the first Amphibia, from a Greek word meaning "those with two lives."

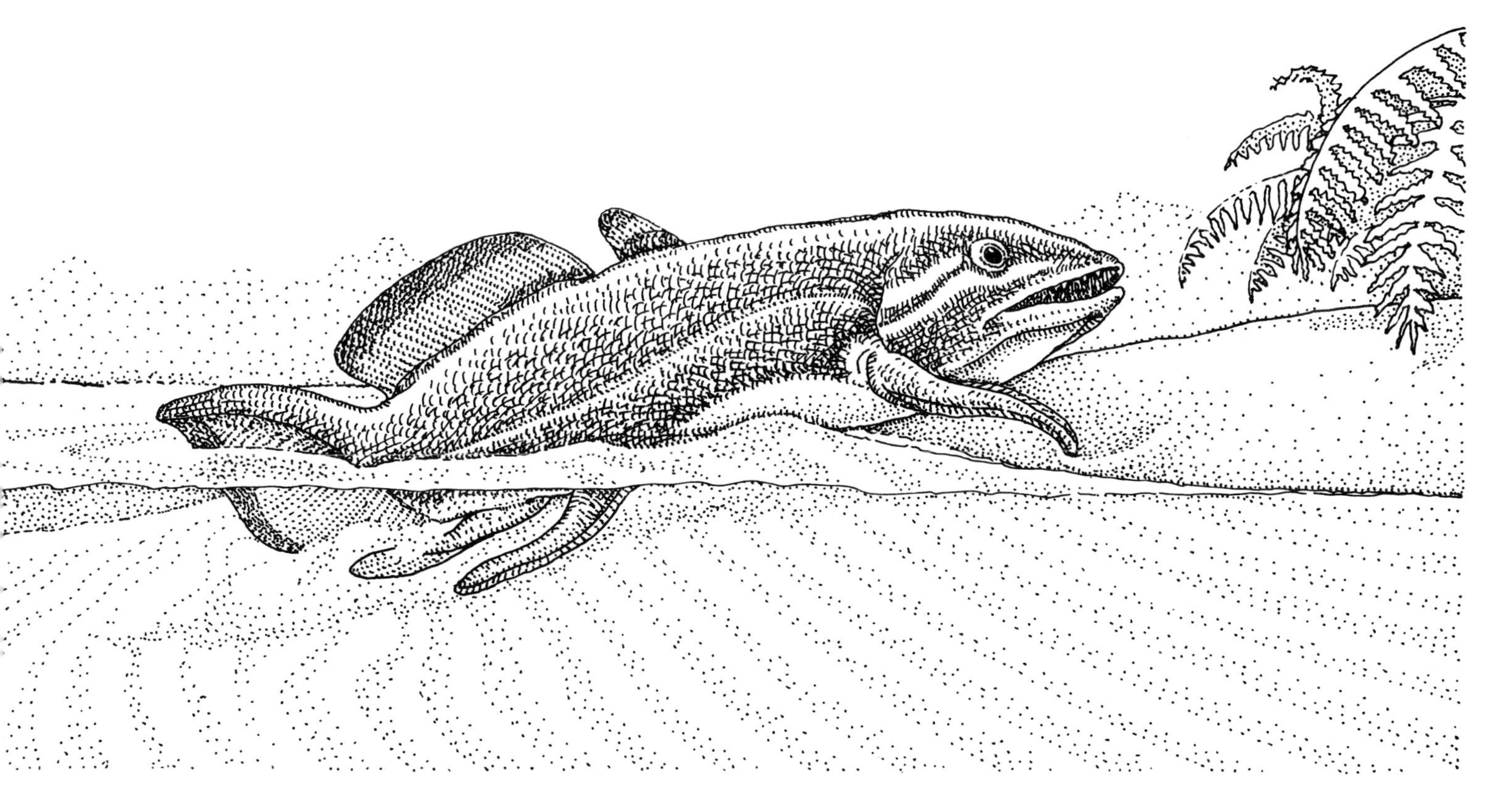

needed their lungs and strong fins not to leave the water, but to seek water. These adaptations are not forward-looking but conservative, backward-looking attributes designed to keep the fishes in good water.

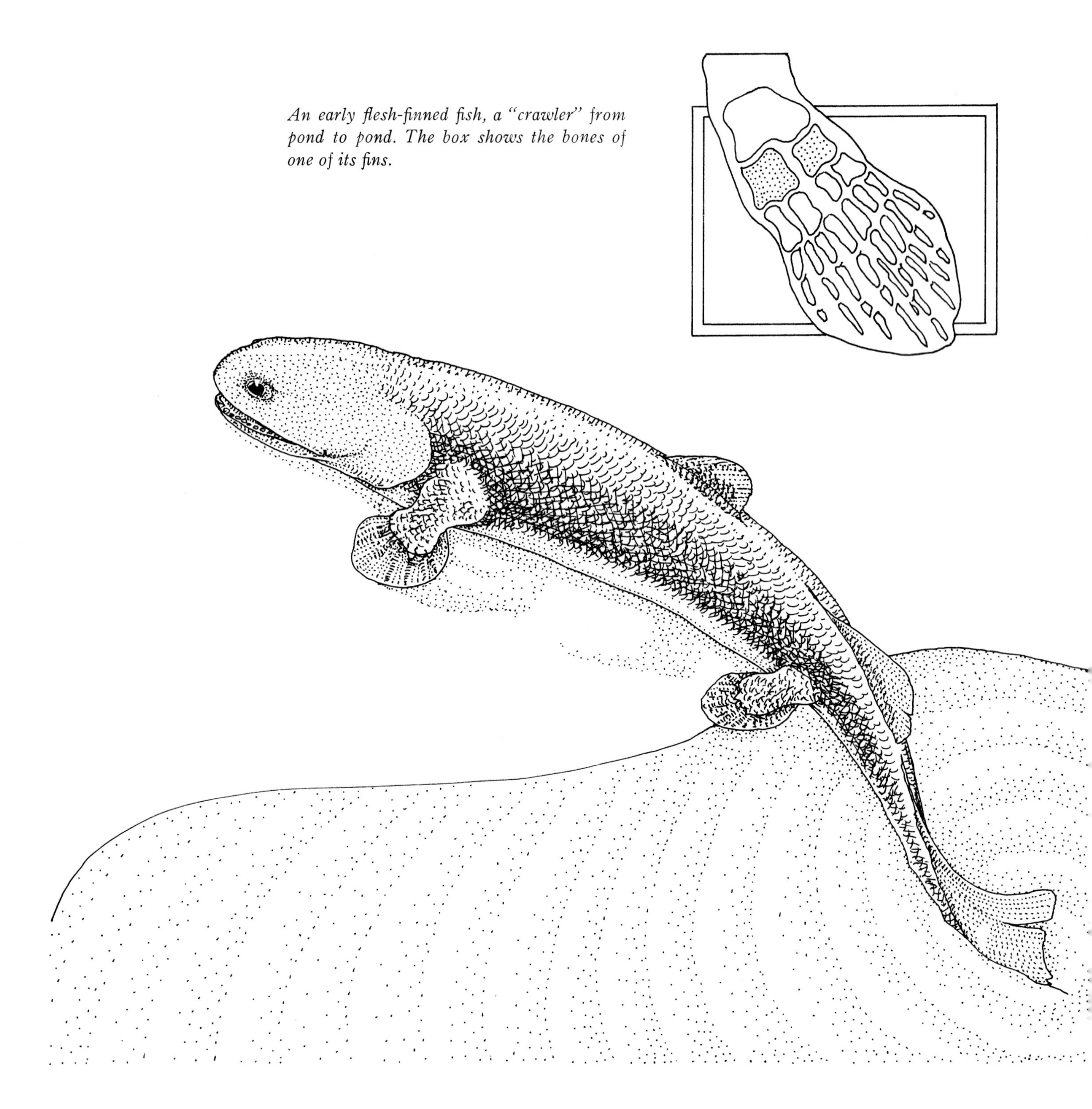

An early flesh-finned fish, a "crawler" from pond to pond. The box shows the bones of one of its fins.

Early amphibians were confined for most of their lives to the water, where they lived and loved much as did their fishy ancestors. When life became too hard, their stubby legs simply served to carry them from one stinking, stagnant body of water to another. Their lungs sustained them during these periods of transition; gills were still used for water-breathing. Their skin was covered with a layer

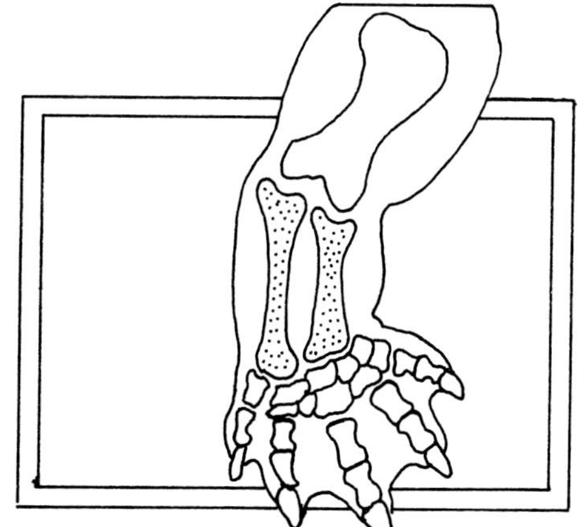

An early amphibian and the bones of one of its feet. The shaded bones evolved from the similar (homologous) bones of the fish fin (opposite page), *also shaded.*

of heavy mucous, passed along from their ancestors, and it helped them retain their own internal moisture when out of the water.

These first amphibians were the most successful vertebrates in the swampy pools in which they evolved. And, as always happens with successful animals, the amphibians overpopulated their econiches, pressing one another into extinction or the good luck of change.

During their tremendous adaptive radiation, amphibians learned to use almost every food source available in the swamps—and some on dry land, to boot. Recalling that the plants were the first living forms to invade the continents, closely followed by the arthropods, we can envision a time around 400 million years ago when the earth's lowlands were covered with gigantic forests of club mosses and other primitive trees. On this vegetation fed an ever-increasing host of insects, many of which were themselves of rather large size. Some early dragonflies had wings that spanned one meter. The smaller insects served as food for their larger relatives and for any other beings fast enough to catch them. In the swamps, the growing number of amphibians forced a few of the smaller, more active varieties to spend more time on land. And on land, insects were the most nourishing food available.

Now, in order to catch insects, one has to be fairly quick on one's feet. Thus only the tiniest and lightest amphibians became land-dwelling insect hunters. Being small themselves, however, these amphibians were in perpetual danger from their larger cousins whenever they reentered the water. These pressures ultimately created a new type of amphibian: one whose entire life (almost) was spent on land, and whose small size and more efficient legs produced a quick-scrambling insect chaser, a new kind of vertebrate. Still, life on land subjected these animals to a variety of new pressures.

The amphibians, bound to water by the nature of their reproduction, nonetheless founded an adaptive radiation of their own. As we will see, it was a radiation very important to us, their remote descendants.

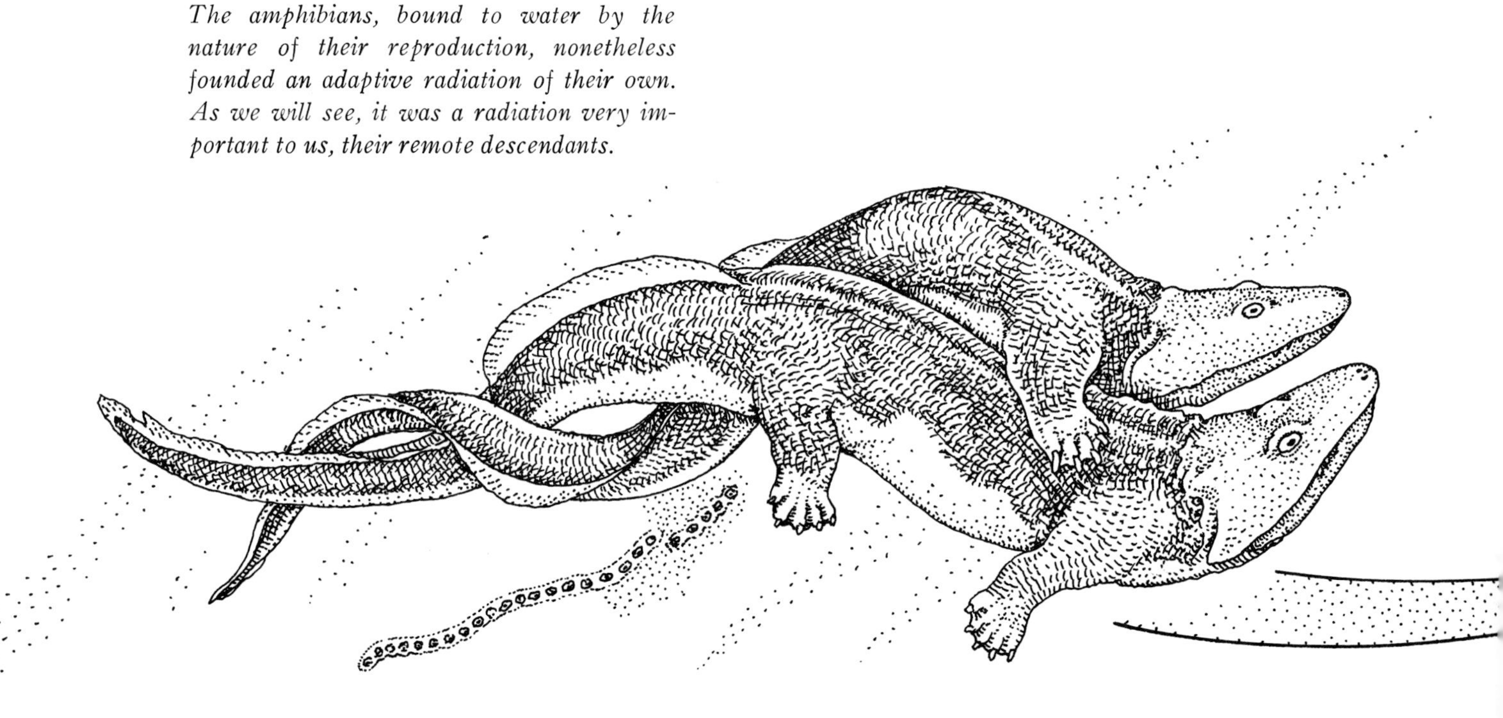

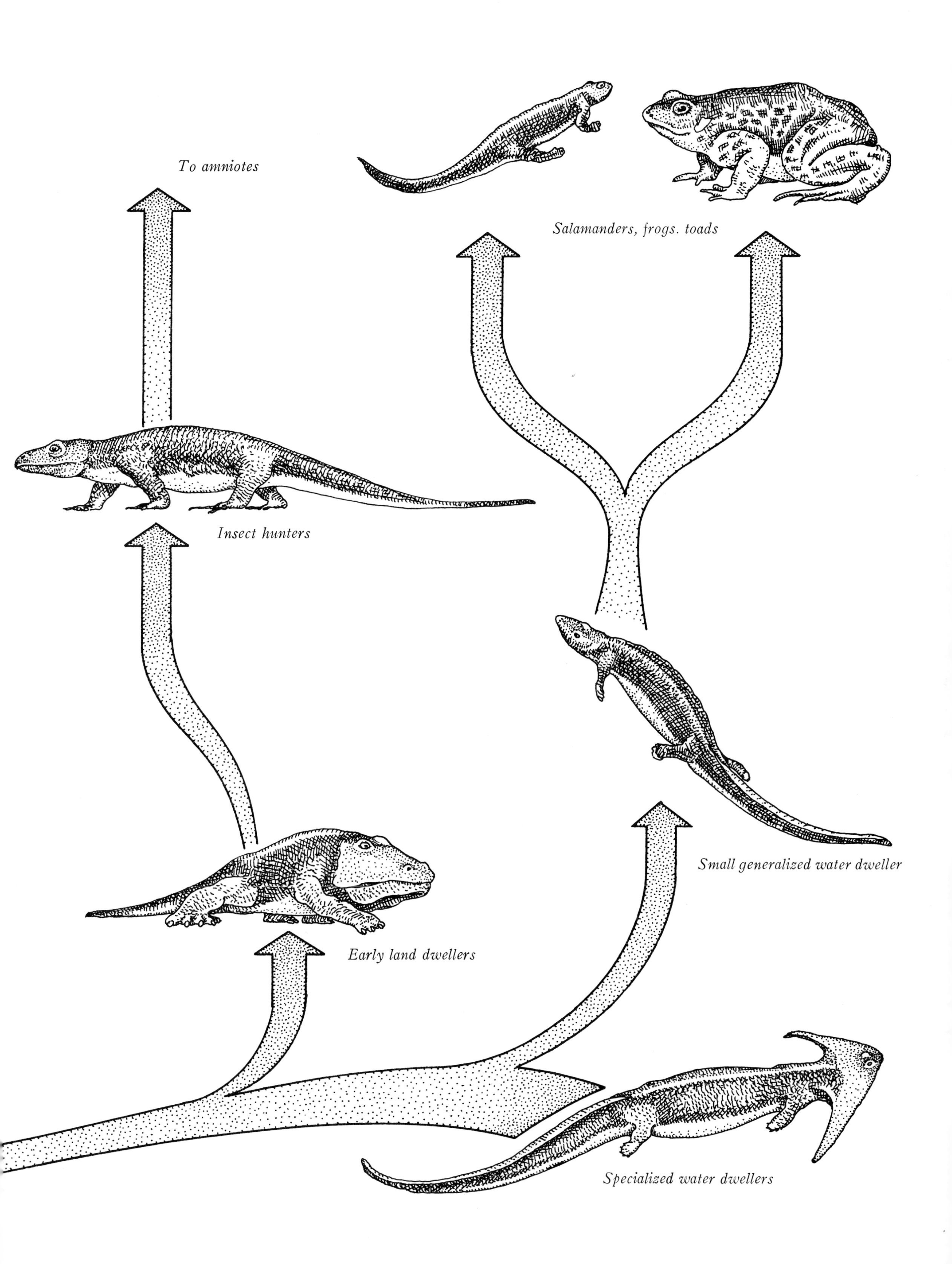
To amniotes
Salamanders, frogs. toads
Insect hunters
Small generalized water dweller
Early land dwellers
Specialized water dwellers

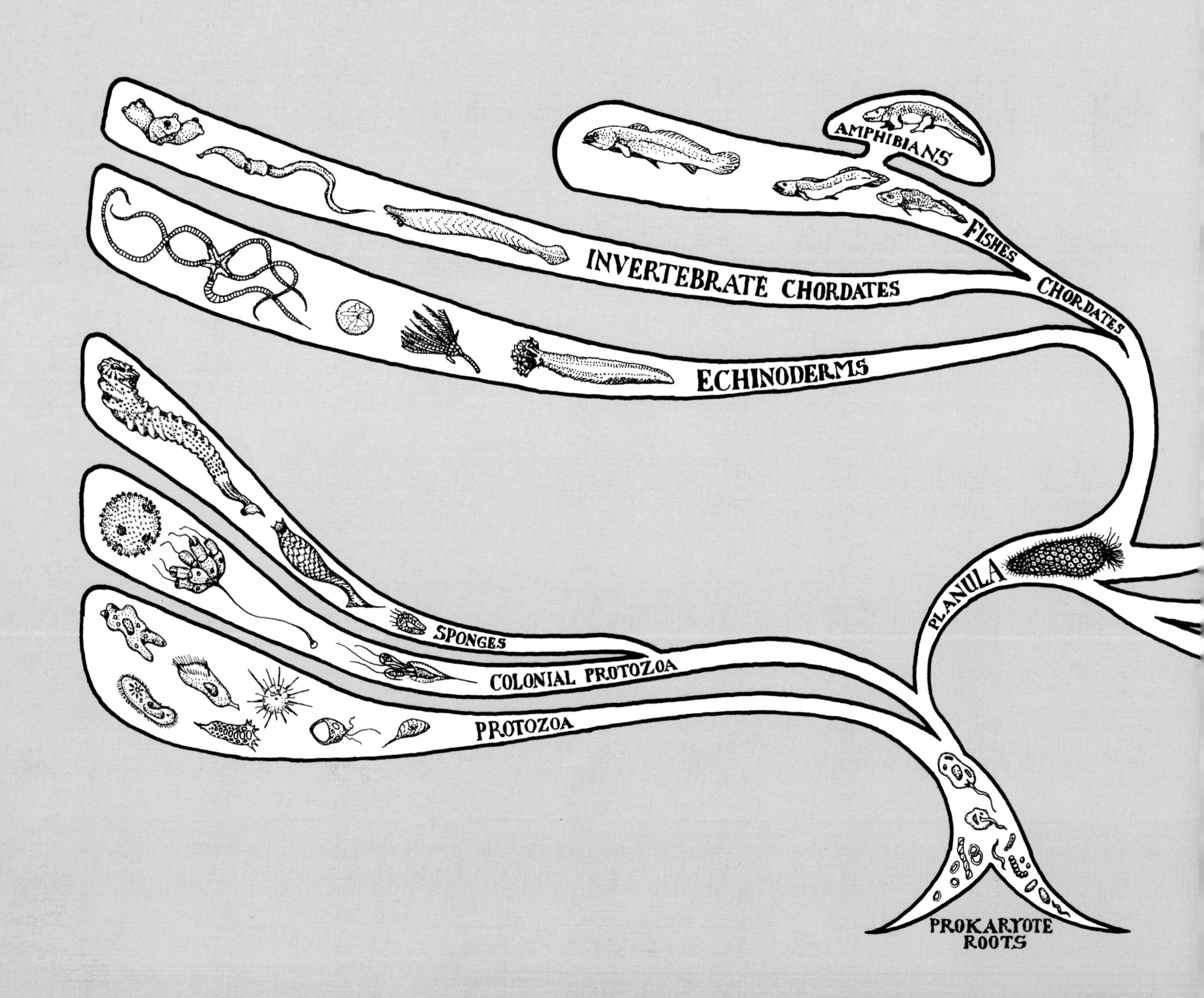
AMPHIBIANS
FISHES
INVERTEBRATE CHORDATES
CHORDATES
ECHINODERMS
PLANULA
SPONGES
COLONIAL PROTOZOA
PROTOZOA
PROKARYOTE ROOTS

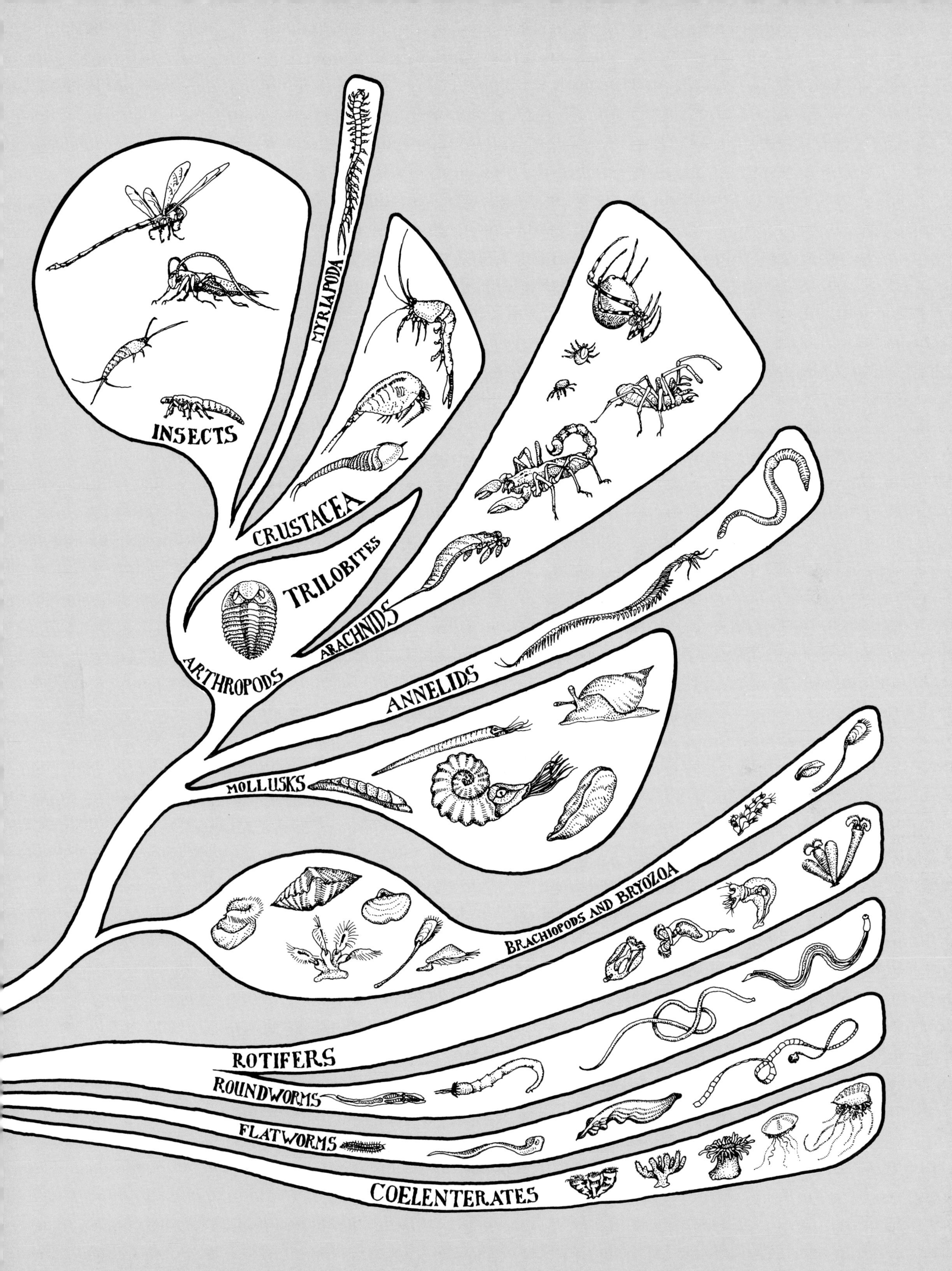
INSECTS
MYRIAPODA
CRUSTACEA
TRILOBITES
ARACHNIDS
ARTHROPODS
ANNELIDS
MOLLUSKS
BRACHIOPODS AND BRYOZOA
ROTIFERS
ROUNDWORMS
FLATWORMS
COELENTERATES

Those amphibians that retained a typically moist, slimy skin were liable to dry up in sunny weather. Mutants appeared with leathery, waterproof skin that protected the body's water, and they were better able to survive. Again, being amphibians still, these early land-lubbers laid their eggs in water. Unfortunately, whenever they did so, they were liable to be eaten by their bigger neighbors, and so were their eggs.

Mutations occurred that eventually led to internal fertilization, in which the sperm of the male traveled directly from his body into the female's. This was an improvement over the method for water-dwelling amphibians, where the sperm had to pass through open water in order to reach eggs the female had previously deposited. The egg formation process itself became more complex, as natural selection tended to favor mutations that overcame the vulnerability of amphibian eggs and their young.

Some lucky mutations eventually produced the amnion, a thin sac enclosing both the fertilized egg and a small body of moisture, a kind of "portable pond" (amniotic fluid). Within this egg a youngster might pass the early, vulnerable stage of its life. From such an egg the young animal could hatch as a fully land-dwelling miniature of its adult form, instead of the helpless fishlike larva characteristic of amphibians. Later in its evolution, the amniote egg would be enclosed by a shell and contain a nourishing yolk. This kind of amniote egg is characteristic of all true land-dwelling vertebrates, the reptiles, the birds, and the mammals. All of them are therefore called "amniotes." The appearance of the amnion around 310 million years ago liberated its possessors and permitted them to move unlimited distances from bodies of freshwater.

The first amniotes spread nearly unopposed across the warmer parts of the world's landmass. (At that time, most of the continents were concentrated in one supercontinent, which is dubbed Pangaea, "all land.") These first amniotes were like lizards in appearance,

The contrast between a typical amphibian egg (opposite page, left) *and a typical amniote egg* (opposite page, right). *In the amniote egg, the embryo floats in its own little pond (amniotic fluid, shaded), whereas the amphibian embryo is unprotected. A fishlike larval amphibian with external gills emerges from the amphibian egg (lifesize eggs shown in box); a fully active newborn emerges from the shelled amniote egg.*

with long, low bodies and short legs. Scientists traditionally group them in an order Cotylosauria ("stem lizards") of the class Reptilia ("crawlers") of primitive amniotes.

Class Reptilia was officially named by the great naturalist Karl von Linné in 1737. Von Linné erected a system of classification of living forms, the *Systema Naturae,* which has remained the basis for the modern science of taxonomy, the grouping of organisms according to their lines of descent or relationship. When he named the reptiles, von Linné was thinking of those living amniotes that creep low upon the ground—the snakes, lizards, turtles, and crocodilians—and it was to these modern animals that he applied the term "crawlers."

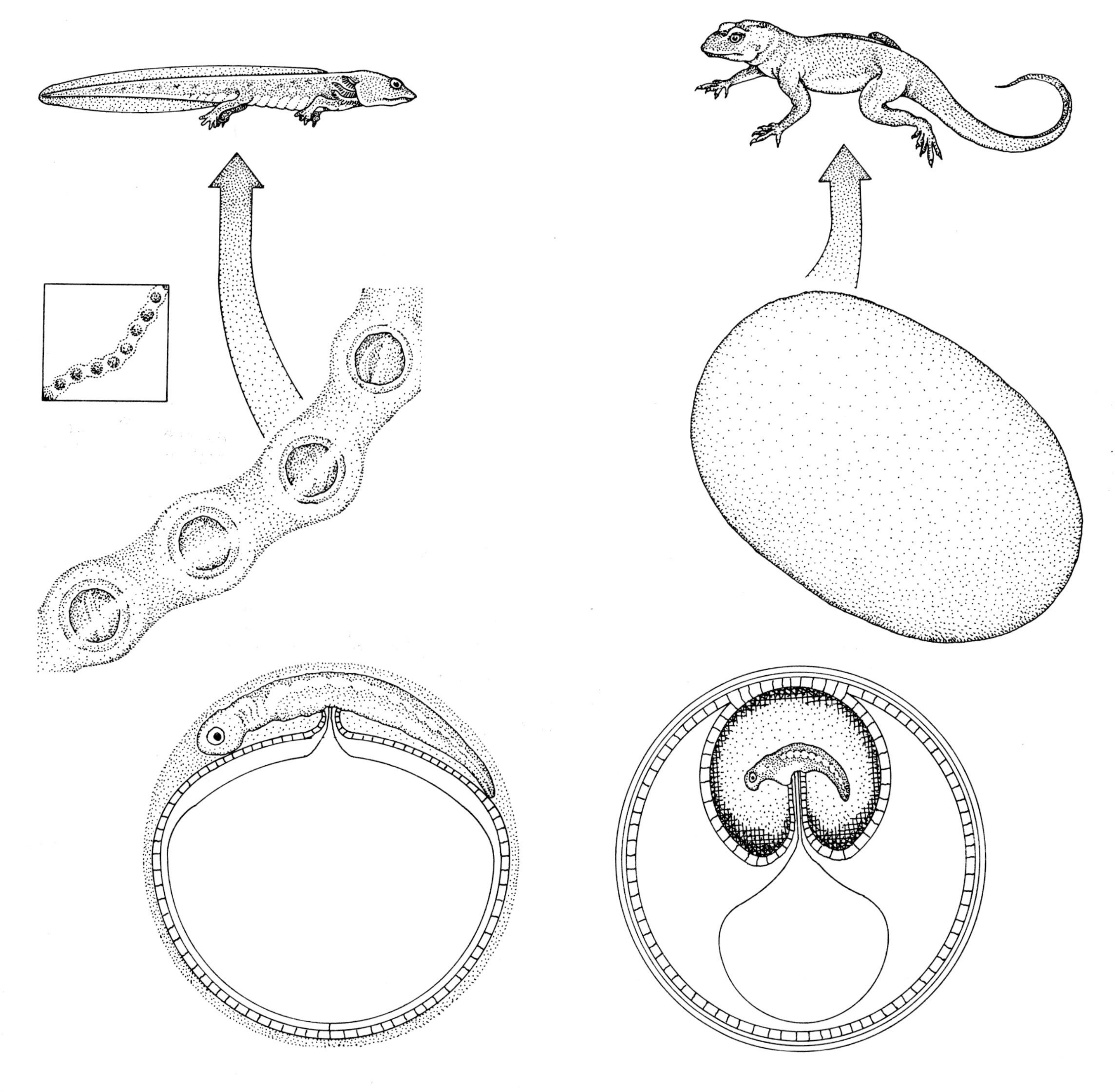

Since von Linné's time, however, hundreds of fossils of extinct amniotes have been found that are in many ways quite unlike anything alive today. Because all amniotes seem to have evolved from forms that resemble living reptiles, many extinct amniotes were and still are routinely included among the Reptilia, even though they were not crawlers at all. Dinosaurs, for instance, are usually called reptiles. Yet if one of these long-gone creatures had appeared alive before von Linné, he would probably no more have called it a reptilian "crawler" than he would so have called a bird. This sort of misnomer becomes a problem in the study of animal evolution, for when we call an extinct animal a reptile, we still tend to *think* of it as a crawler, even though it may actually have been a furry flyer or a feathered two-legged runner. To avoid this confusion, we shall call fossil amniotes names other than "reptiles," except for those (like lizards) whose descendants are actually true crawlers.

A riverbed around 310 million years ago. The small insect-hunting amniote Hylonomus *is approached from left by a two-meter-long amniote,* Limnoscelis, *the largest and*

The first amniotes, the cotylosaurs, were limited (as are all true reptiles) by their inability to control their body temperatures internally. This condition is known as ectothermy, "heating from without," and we can see it today in snakes, lizards, and turtles. Ectothermic animals must spend a good deal of time moving from sun to shade (to cool off) and back again to sun (to heat up) in order to maintain the body temperature at which they function best—generally a warm one; few if any reptiles are found in cold climates. So it was with the primitive cotylosaurs, which seem to have been restricted to lowland areas near the Equator, where the climate was (and still is) warm for most of the year.

meanest vertebrate on land. Hylonomus *was about a meter long; its fossils are found in tree trunks. These animals controlled their body temperatures by seeking shade or water when hot, sunlight when cold.*

The cotylosaurs were so good at catching insects, amphibians, and each other (the first amniotes were all carnivores) that their numbers grew, and they quickly began to intrude upon one another. Cotylosaurs with traits that conferred special advantages for survival—that fitted them for a particular food or a habitat where they would encounter fewer predators, for instance—tended to survive this competition and reproduce. As different characteristics strengthened and developed in different locations, the group as a whole became more diverse. Early in their history we find cotylosaurian amniotes producing a number of distinct lines of evolution that would enable their descendants to invade a variety of new niches and further consolidate their hold on the world.

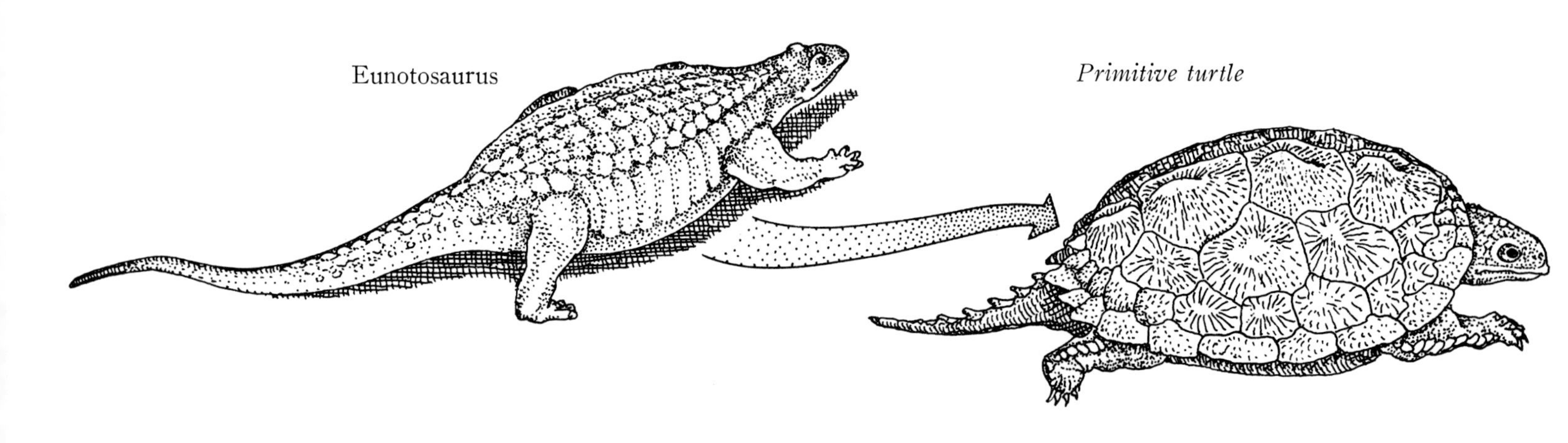

The initial cotylosaur adaptive radiation produced such diverse amniote forms as early herbivores (top); *armored forms* (above) *from which descended the modern turtles, themselves little changed over the intervening hundreds of millions of years; real modern reptiles, such as the lizard, or squamate* (opposite page, above); *early fish-eating aquatic forms* (opposite page, below).

Many of them, capitalizing on their lowslung reptilian form and rather slow way of life, remained unchanged over long periods of time. Among their descendants are the squamates ("scaled ones"), the snakes and lizards with whom we share the land today. Turtles and tortoises are also descendants of the conservative cotylosaurs. With the development of their familiar body armor, turtles and tortoises seem to have settled early for a lifestyle like that of the very first amniotes, having for the most part remained associated with bodies of water.

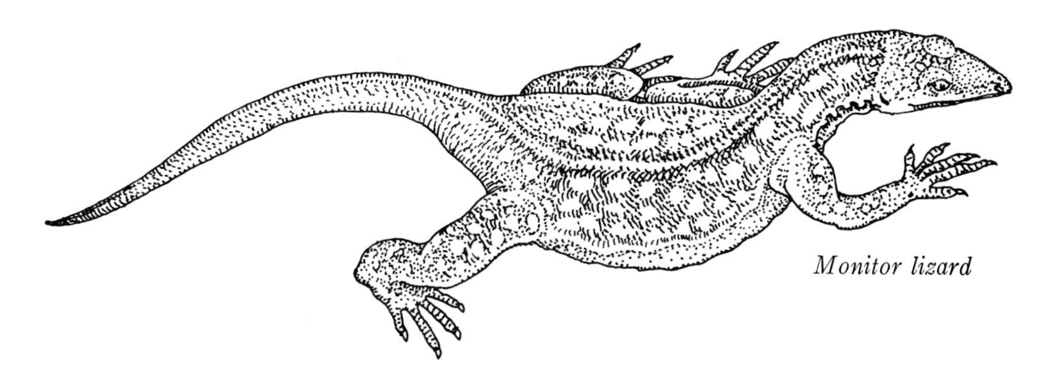

Monitor lizard

Other cotylosaurs, however, went on to mighty deeds. Some seem to have returned to a watery habitat, becoming swimmers and eating fishes and amphibians. So well did some cotylosaurs adapt to aquatic life that they reinvaded the oceans to take up life as large, fast-swimming, air-breathing predators (a niche now inhabited by modern toothed whales and seals). The cotylosaurs in freshwater took a deadly toll on the amphibians. As a result, amphibians became extinct or highly reclusive. In fact, of the many great amphibians that once ruled the swamps, only the little frogs, toads, and salamanders now remain in a vertebrate world dominated by the more advanced amniotes.

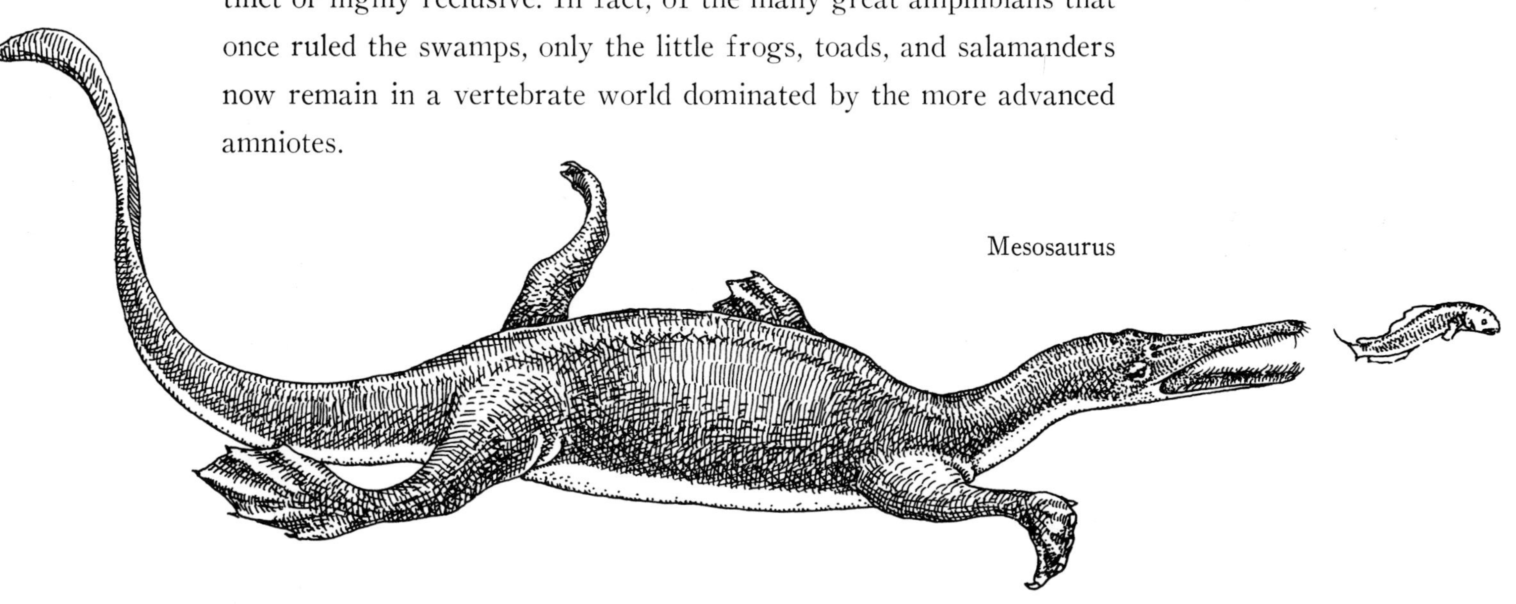

Mesosaurus

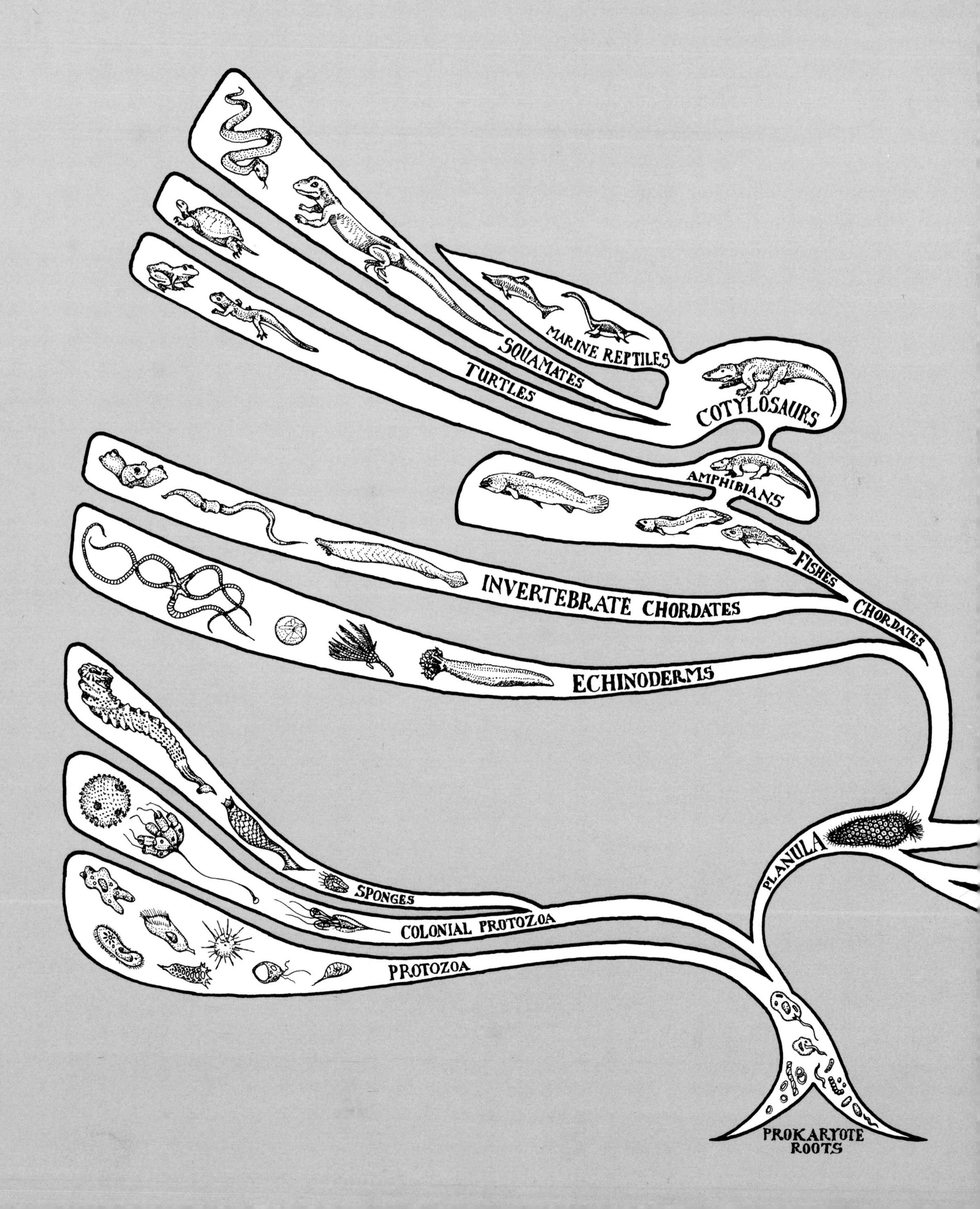

MARINE REPTILES
SQUAMATES
TURTLES
COTYLOSAURS
AMPHIBIANS
FISHES
INVERTEBRATE CHORDATES
CHORDATES
ECHINODERMS
PLANULA
SPONGES
COLONIAL PROTOZOA
PROTOZOA
PROKARYOTE ROOTS

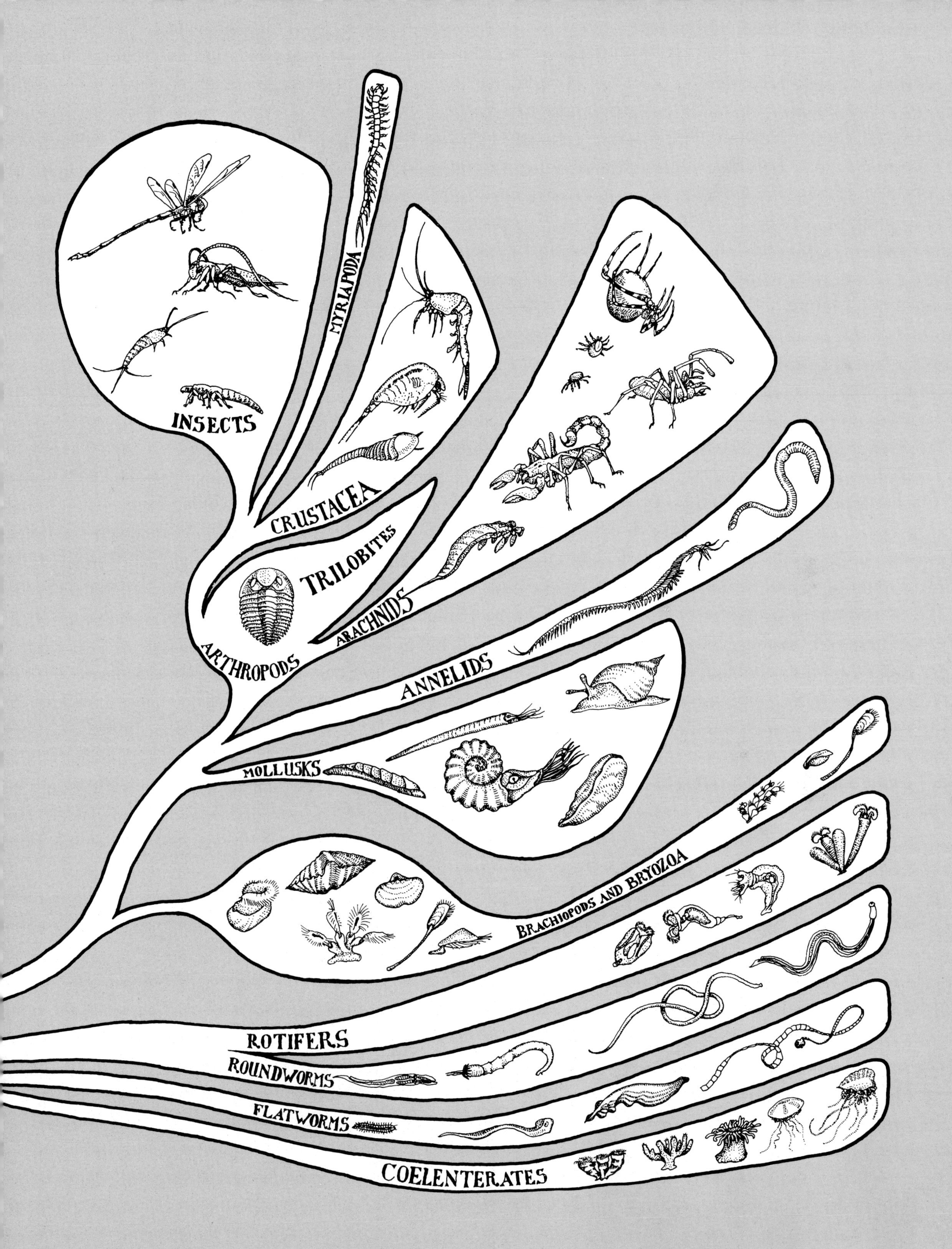

INSECTS
MYRIAPODA
CRUSTACEA
TRILOBITES
ARACHNIDS
ARTHROPODS
ANNELIDS
MOLLUSKS
BRACHIOPODS AND BRYOZOA
ROTIFERS
ROUNDWORMS
FLATWORMS
COELENTERATES

Pelycosaurs, early swamp-dwelling synapsids. (left) *A small, active form,* Haptodon, *sits atop the skull of a larger relative.* (right) *The more specialized* Ophiacodon *ate fish*

Other early cotylosaurian amniotes took to life on land with greater verve than had most of their vertebrate contemporaries. They continued to scamper after insects and to perfect their speed and agility. This high-speed lifestyle required a high level of energy, a requirement that these amniotes met by eating more insects, and maybe even each other. This intensified the pressure on insects, causing them to evolve. Natural selection eventually produced new orders of insects with lifestyles based on smaller size, greater flying speed, and other defenses such as camouflage, bitter taste, tougher exoskeletons. And as the insects evolved, the amniotes that ate them coevolved: they developed better legs and more efficient jaws with which to catch their wiggly prey.

One of the most significant advances among certain cotylosaurian amniotes was the evolution of openings for big jaw muscles that permitted a wider mouth gape and quicker biting. An early example of this trend is found among synapsid ("fused-arched") amniotes. The name refers to the shape of the arched opening for the jaw muscles

and amphibians. Casea, *the sluggish plant-eating giant behind* Ophiacodon, *was some three meters long.*

in their skulls. The large jaw muscles that closed the jaws were attached to the inside of the rear of the skull, and the arch permitted the muscles more room and thus more power. On account of the selective advantages their jaws gave them, the first synapsids diversified into, on the one hand, conservative swamp dwellers that ate fish and even plants, an unusual econiche among early amniotes; and, on the other hand, still faster insect eaters that went even further into the uplands as successful amniotes pressed on one another and used up the lowland econiches.

Synapsids were from the start very advanced for their times. Even the conservative swamp-dwelling varieties were energetic, eating more food and moving more efficiently than the cotylosaurs from which they were descended. Still, they were limited by their lack of internal temperature control, which confined them to moving about in warm places during warm hours of the day. In addition, they could not grow especially large, because they controlled their temperatures by taking up or giving off heat through their body surfaces.

In the changing conditions of day and night on land, only comparatively small animals can perform this heat exchange efficiently. As body volume increases, so does body surface area. But the volume increases much faster than does the surface area; in other words, with a given increase in volume, there is not a sufficient increase in surface area to permit heat to be released or absorbed quickly enough to accommodate the animal's needs.

Still, large size offers a selective advantage in evading predators and storing body heat, and some synapsids did evolve toward larger size. Many of the swamp-dwelling variety apparently overcame the

Solar-heated pelycosaurs of the genus Dimetrodon, *"two-measure teeth." The animal on the left feels too warm, and has turned its sail to parallel the sun's rays in order to radiate some of its body's heat. The one on the right feels too cool and has turned its sail across*

problem of the surface-to-volume ratio by means of large "sails" they possessed. These sails were finlike extensions of the vertebrae that supported thin membranes of skin richly supplied with blood vessels. Such fins greatly increased body surface area. By regulating the flow of blood through them, these animals could employ the fins either as solar heaters, picking up the sun's heat, or as radiators, releasing body heat into the surrounding air. This was one of the earliest vertebrate steps toward an efficient temperature control system for large animals.

the sun's rays to soak up heat. Blood flowing through the sail will warm up and return to the creature's muscles, permitting him to lunge at his rival. Dimetrodon *was, as its appearance indicates, a powerful predator.*

The smallest synapsids, unencumbered by surface-to-volume problems, continued their expansion into the uplands. To escape cannibalistic relatives, they responded with the evolution of fast-running legs that could lift their bodies off the ground. This, in turn, required of them more efficient musculature and an increased need for energy. More energy meant eating more insects. Such pressure gradually forced these littlest synapsids to abandon the old reptilian, or crawling, body posture in favor of one that held the body ever further off the ground for better speed. This required more active muscles and thus ever greater appetites.

A growing appetite is the result of an increased need for energy. In order for that energy to be available to the body quickly, food must be broken up into small pieces, so that digestive juices can reach all of the food as soon as possible. A snake, a reptile that leads a slow, easy life, eats only once in a great while and swallows its prey whole. Then it goes into hiding while its digestive juices gradually dissolve the animal. The fast little synapsids, on the other hand, were

The progress of the synapsid line toward more efficient locomotion and higher rates of activity was marked by improvement in their jaws and teeth. The "apse," or opening in the skull for powerful jaw muscles, is marked by an A in each drawing. As the musculature of the jaw becomes better adapted for chewing, and the teeth more differentiated, the apse grows in size. With this progress, the animal's gait also improves.

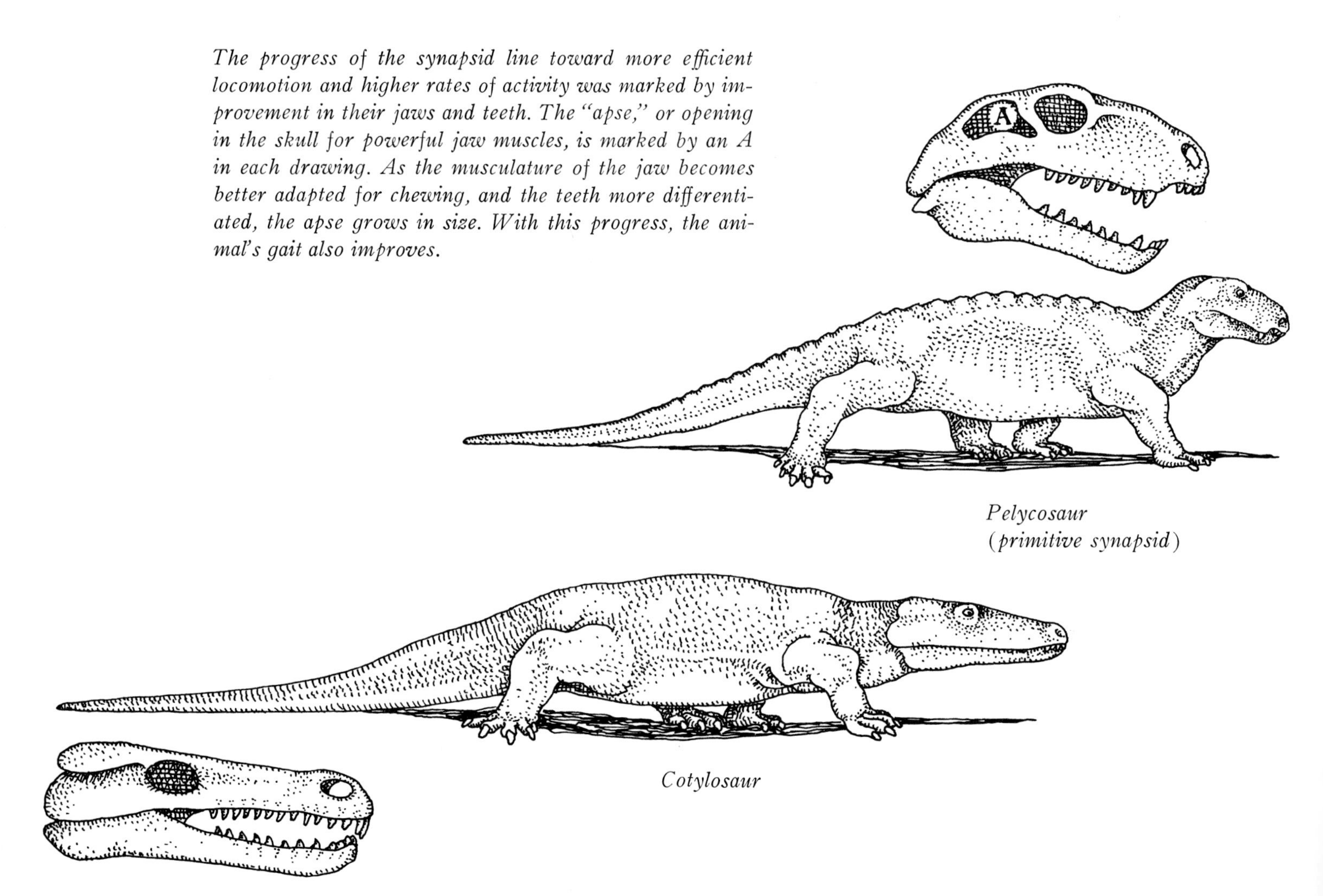

Pelycosaur (primitive synapsid)

Cotylosaur

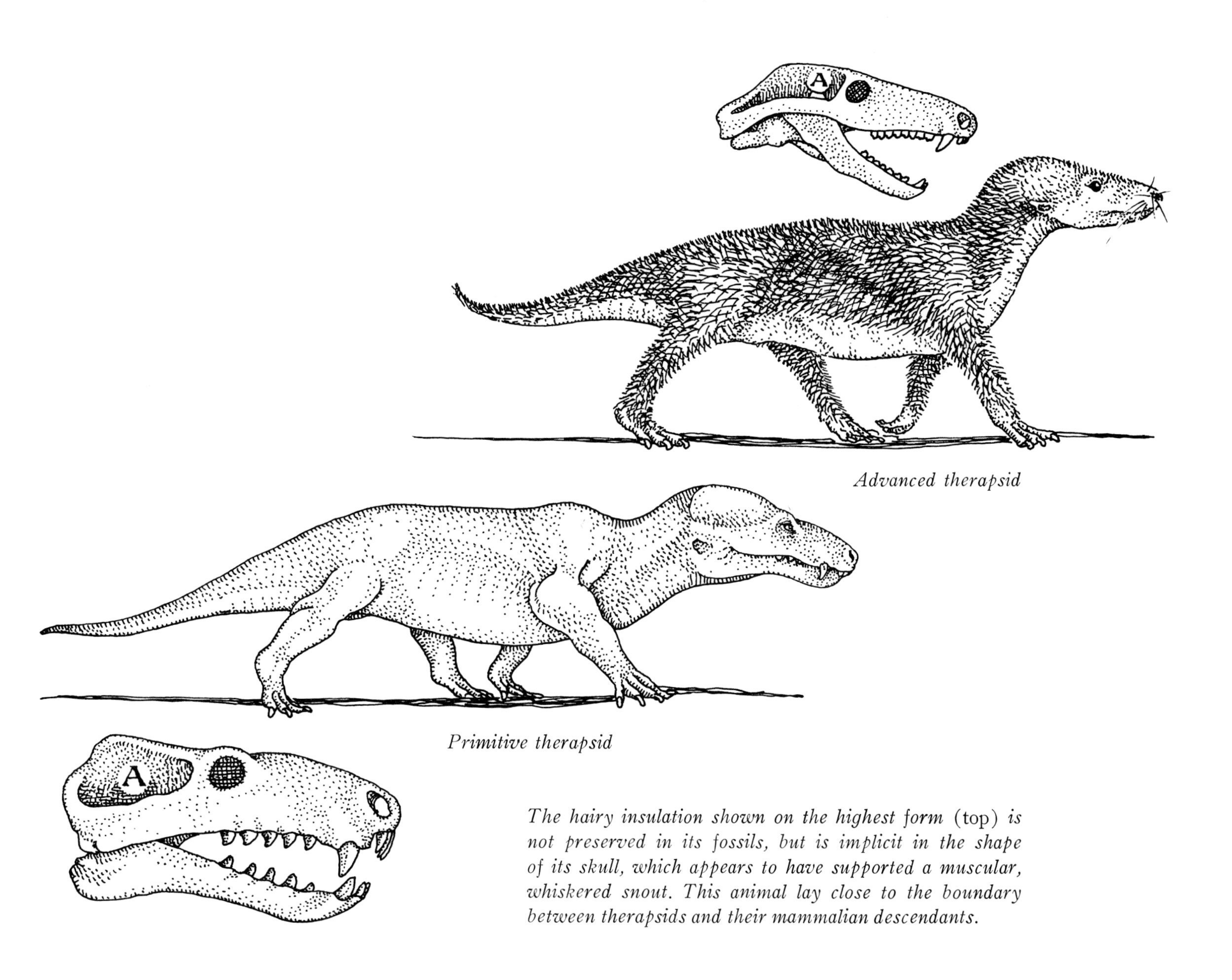

Advanced therapsid

Primitive therapsid

The hairy insulation shown on the highest form (top) *is not preserved in its fossils, but is implicit in the shape of its skull, which appears to have supported a muscular, whiskered snout. This animal lay close to the boundary between therapsids and their mammalian descendants.*

forced to eat all the time. They couldn't wait for slow digestion to work on a great lump of food. Those that had some mechanism for breaking food into pieces gained an energy advantage over those that didn't. The result of this selective pressure was the evolution of teeth, which, from the simple peglike structures in fishes, became instruments of chewing in these small fast synapsids. These were the first real teeth to be found among vertebrates.

As synapsids continued to invade uplands, they came to possess a variety of teeth: some for nipping, which we call incisors; some for cutting or crushing their food, which we call molars and grinders; and, between these, long piercing tusks for holding and killing their prey. Because these piercing tusks are characteristic of dogs, we call them canine (doglike) teeth. Together these different kinds of teeth permitted synapsids to pioneer a whole new way of life based on fast assimilation of energy, fast motion, and high muscle tone.

These faster, more advanced synapsids, possessing complex teeth and standing high off the ground, were quite a new phenomenon. They spread out over most of the earth's surface, even invading cool areas near the South Pole. Initially, they were all insect eaters, but as they thrived, their populations increased; and as their populations increased, the competition for insect food became more intense. The best synapsid insect catchers held on to the original insect-eating niche in which they had evolved. Other synapsids were apparently successful at adapting to different sources of food, and over many generations became more and more specialized for it.

A therapsid scene in South Africa some 230 million years ago. A tusked member of the plant-eating genus Lystrosaurus *is trapped in a mudhole. While the rest of the* Lystrosaurus *herd look on, predatory* Scymnognathi *with saberlike canines surround him. All*

So thorough was the success with which the advanced synapsids invaded new territories and adapted to new food sources and other environmental conditions that distinct new forms evolved. These were the Therapsida, or "mammal-like ones," so called because their teeth and other skeletal characteristics are similar to those of the modern mammals, which ultimately descended from them.

By about 230 million years ago, there were not only small, fast insect-eating therapsids, but great fat plant-eating therapsids that moved across the leafy landscape in mighty herds. On these preyed carnivorous therapsids of large and fearsome aspect, with long sabrelike fangs and great flat heads. The therapsids had conquered the earth for vertebrates, at least for a time.

of these therapsids walk well above the ground on muscular legs. Their high level of muscular activity is made possible by the efficient therapsid chewing mechanisms.

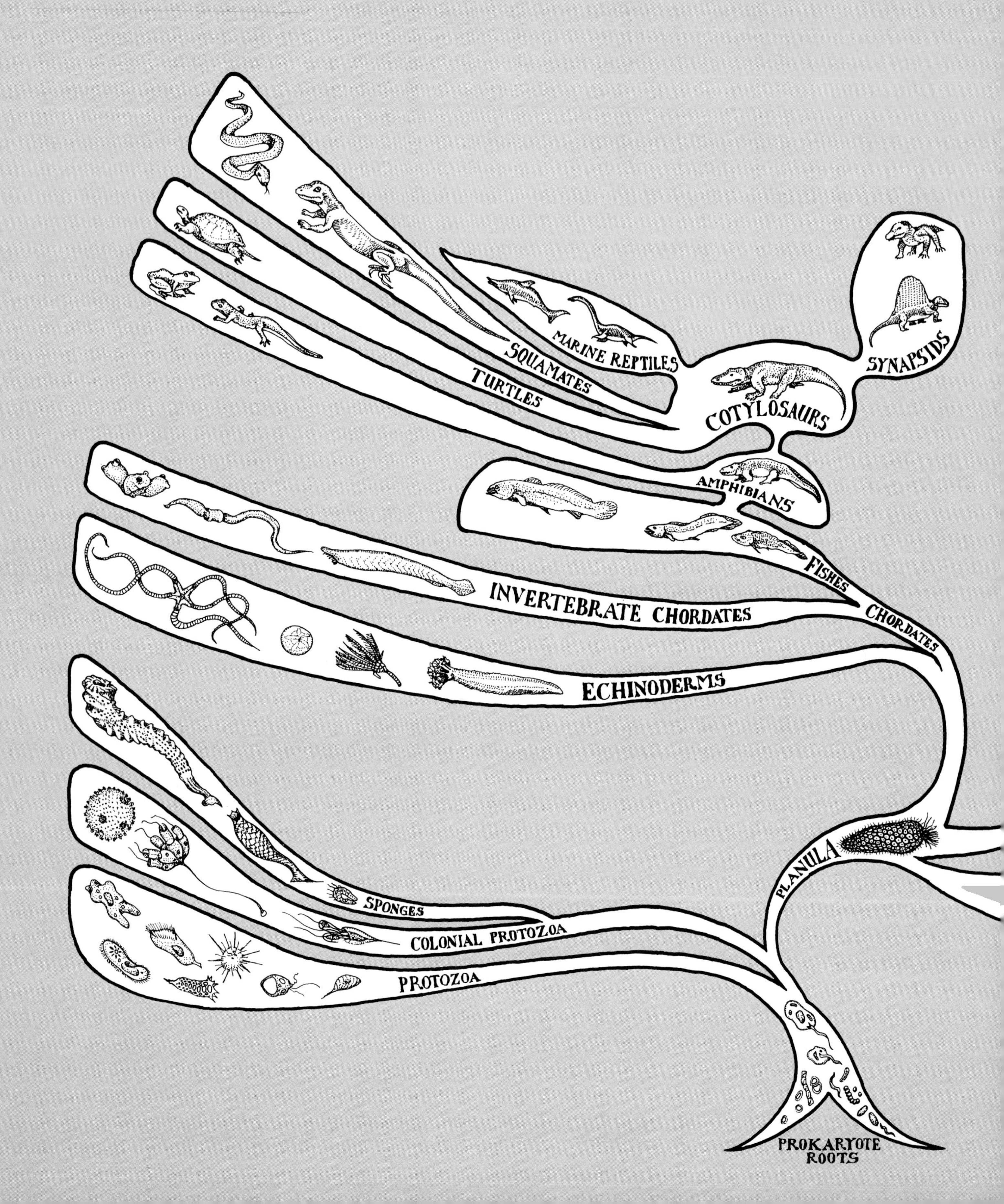

SQUAMATES
MARINE REPTILES
SYNAPSIDS
TURTLES
COTYLOSAURS
AMPHIBIANS
FISHES
INVERTEBRATE CHORDATES
CHORDATES
ECHINODERMS
PLANULA
SPONGES
COLONIAL PROTOZOA
PROTOZOA
PROKARYOTE ROOTS

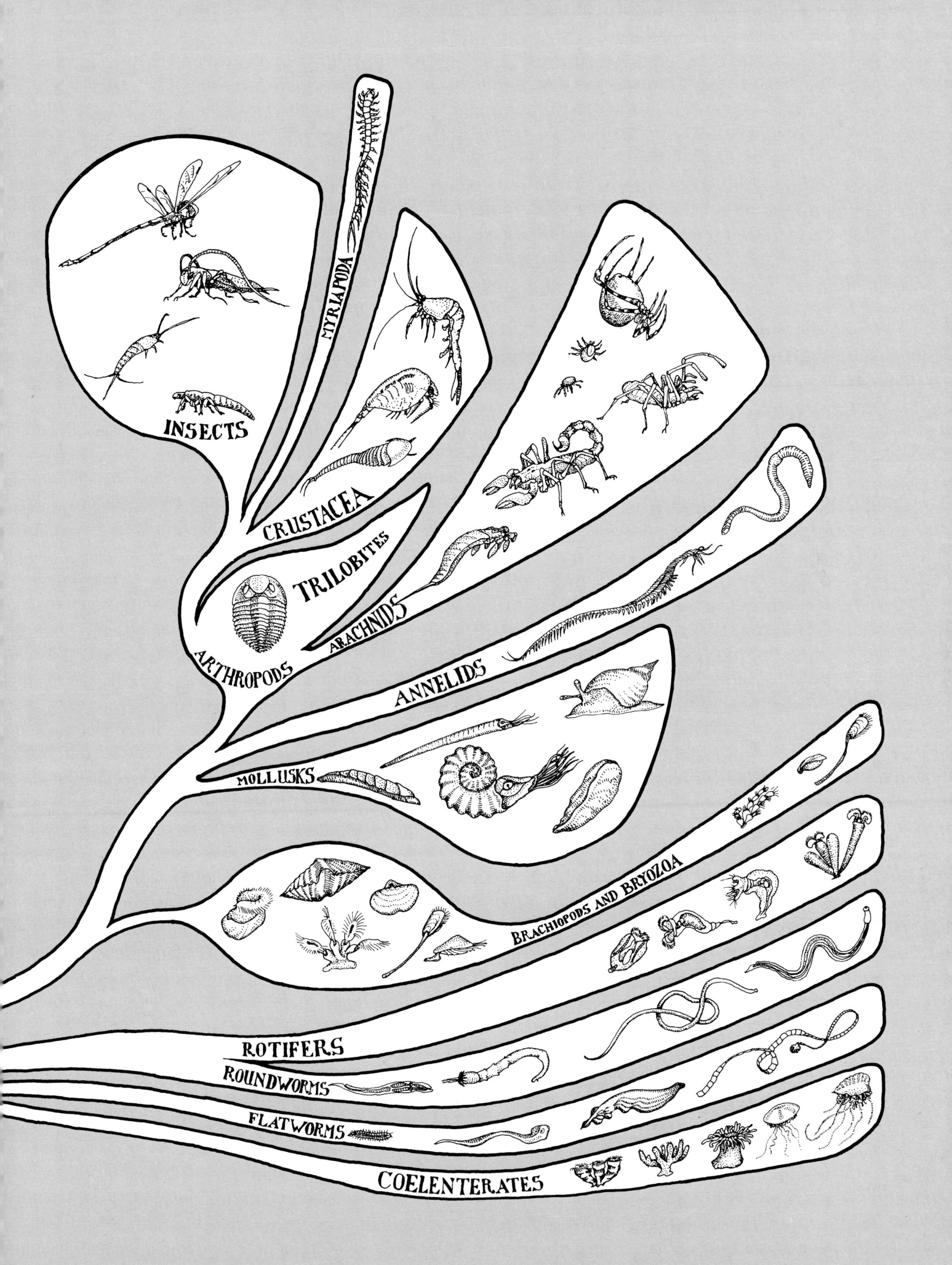

INSECTS
MYRIAPODA
CRUSTACEA
TRILOBITES
ARACHNIDS
ARTHROPODS
ANNELIDS
MOLLUSKS
BRACHIOPODS AND BRYOZOA
ROTIFERS
ROUNDWORMS
FLATWORMS
COELENTERATES

While the therapsids enjoyed their triumph, another group of animals began an experiment with high-speed land dwelling. They originally started out as small alligator-like creatures descended from the early cotylosaurs (the first leathery-skinned amniotes). They made their livings in the swamps, catching fish and the remaining amphibians around 250 million years ago. These swimmers had long tails with which they sculled themselves through the water, and powerful hind legs with which they steered by kicking the bottom, much as do modern crocodiles. Indeed, modern crocodiles are among their descendants.

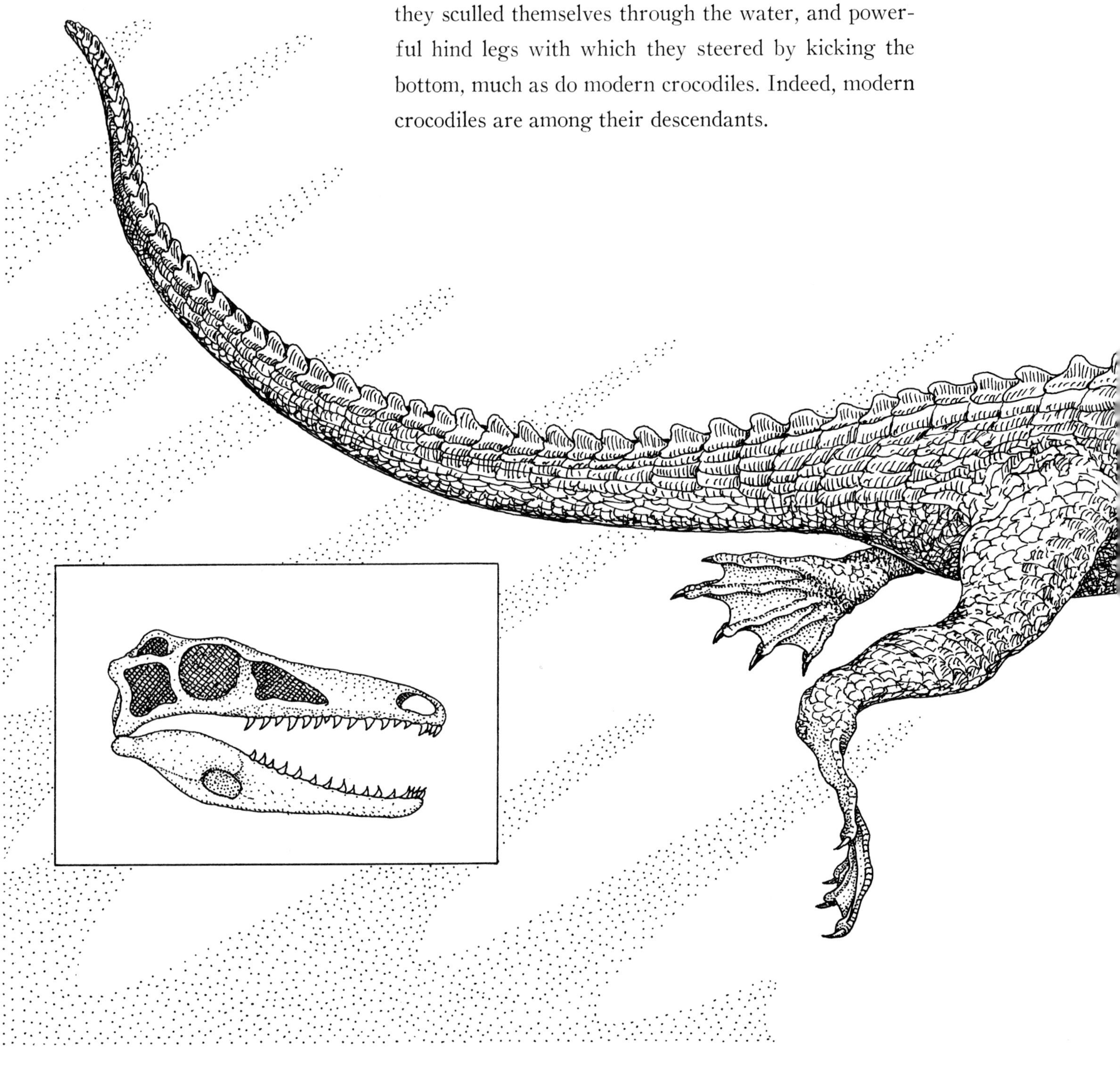

These fish chasers caught their slippery food in long snouts studded with many sharp teeth. Because the teeth were mounted in sockets on the jaws, scientists have named the animals thecodonts, "socket teeth." During the heyday of the therapsids, the thecodonts kept to the swamps, out of reach of the more active animals on shore, living lives rather like those their crocodilian descendants would live. But then the earth's climate began to change, and the thecodonts gave rise to another class of descendants. This group was destined to dominate the world's continents. The humble thecodonts founded the mighty dynasty of Archosauria, "rulers of reptiles," of which the thecodonts themselves were the most primitive members.

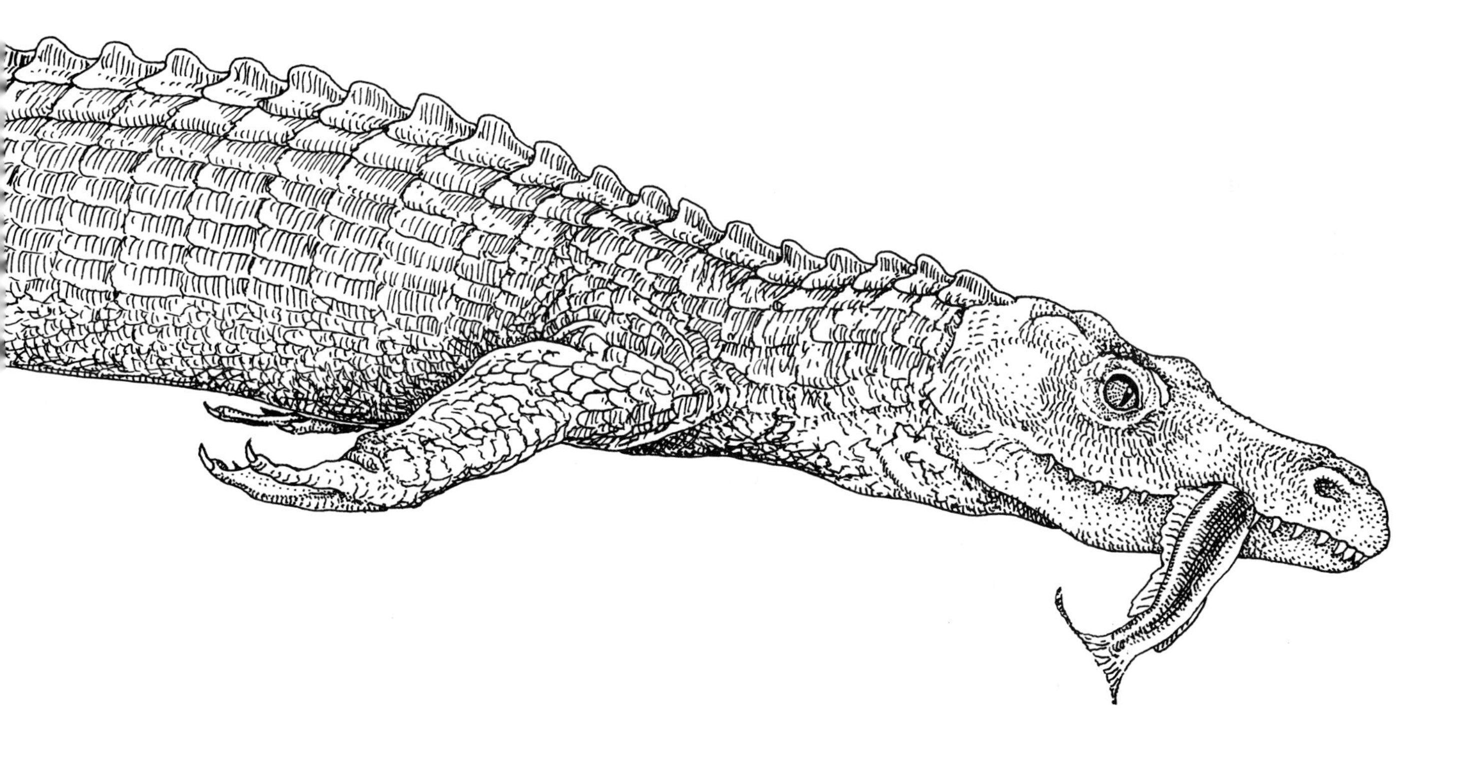

A primitive fish-eating thecodont. The animal looked and lived much like its modern descendants, the crocodiles. (Skull shown in box.)

Along about 230 million years ago, glaciers started to form in the cooler parts of the world, and the Pangaean continent began to dry out as water was frozen into the expanding icecaps. Many swamps disappeared, and shallow-water fish and amphibia became scarce. This of course made life difficult for the crocodile-like, swamp-loving thecodonts. Some of the smaller sorts responded by spending more of their time hunting on land. Here, the only foods suitable for these former fish eaters were small reptiles and insects. Reptiles and insects were able to get about quite quickly by now, since speed had been giving them a selective advantage throughout millions of years of predation. To catch this fast prey, the early land-thecodonts had to be just as fast. Only the swiftest, therefore, survived those hard times, and only the swiftest bore young which in their own turn might survive.

The long hind legs of the thecodonts, originally so useful in swimming, proved useful on land, some legs more so than others. Natural selection weeded out weaker, less adaptive forms, producing a general change in leg structure toward greater speed. Eventually, many thecodonts were able to rise onto their strong hind legs and spring after flying insects. As they ran, their long tails also lifted from the

ground to counterbalance the fore-ends of their bodies. Their fore-legs, temporarily free of walking duty, served to knock down flying prey.

Under this kind of pressure, the slower insects and small reptiles soon disappeared or evolved effective defenses. This, in turn, required of the thecodonts a coevolution. The coevolution took the form of stronger hind legs with larger muscles; better hearts and circulatory systems to carry more energy and oxygen to these legs; sharper vision; better balance and quicker responses to deal with the small, fast beings on which they lived.

Euparkeria, *an early land-dwelling, insect-hunting thecodont.* (left) *The animal at rest.* (right) *In pursuit of an insect, it rises onto its hind legs for a sprint.*

Early on, some thecodont insect eaters took to the trees, chasing their insect prey high into the foliage. As they leaped about from branch to branch, many fell to their deaths. The arboreal (tree-dwelling) life thus exerted pressure on them to defy gravity. Some seem to have been blessed with a lucky mutation that produced flaps of skin which stretched between their foretoes and bodies and acted as parachutes. Gradual improvement of the parachutes resulted in the first flying vertebrates, the pterosaurs, or "winged lizards" (actually winged archosaurs).

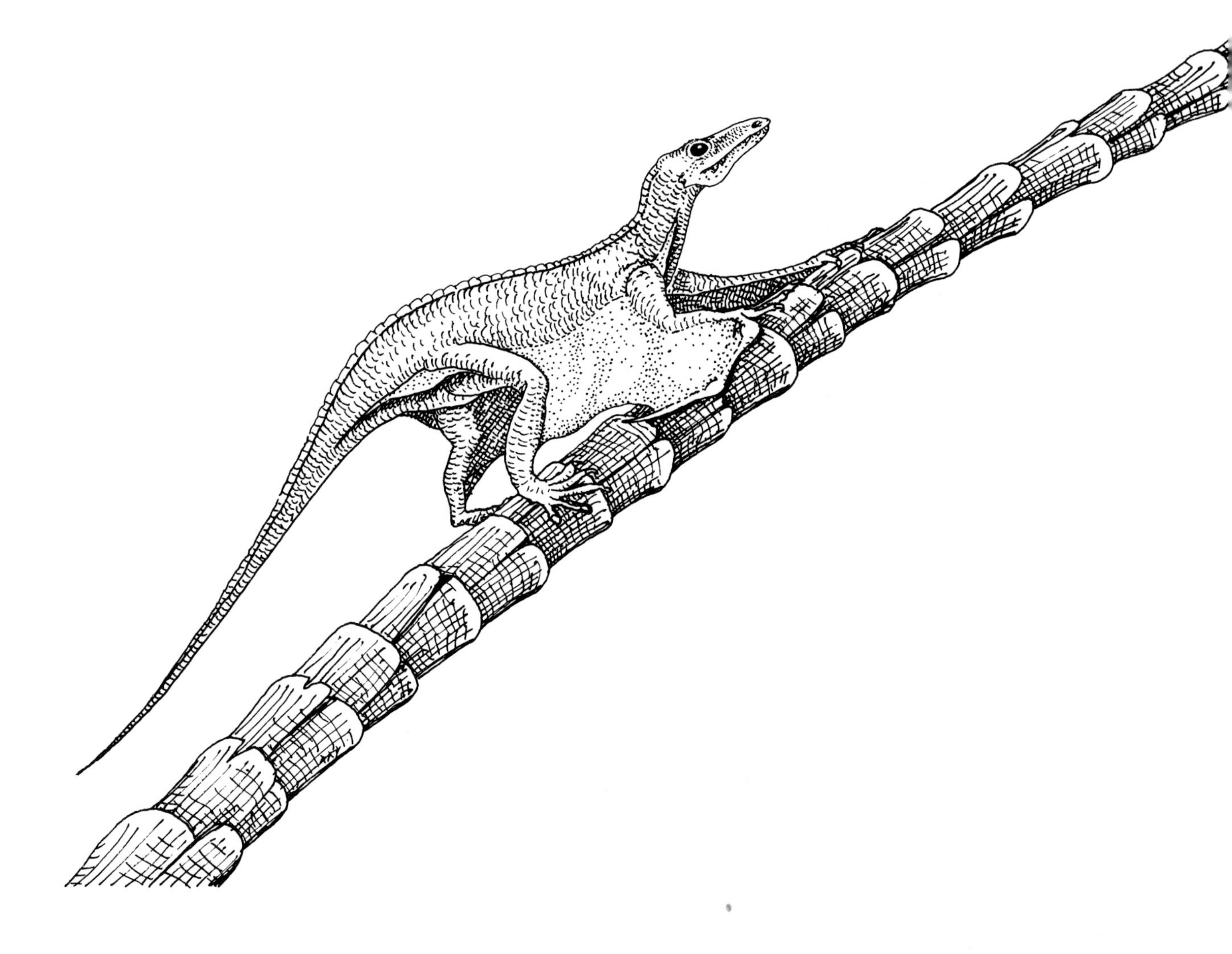

An early (hypothetical) experiment in tree-dwelling among the thecodonts. This climber must have lived rather like a flying squirrel, leaping from one branch to another, aided by the flap of skin stretching between its legs.

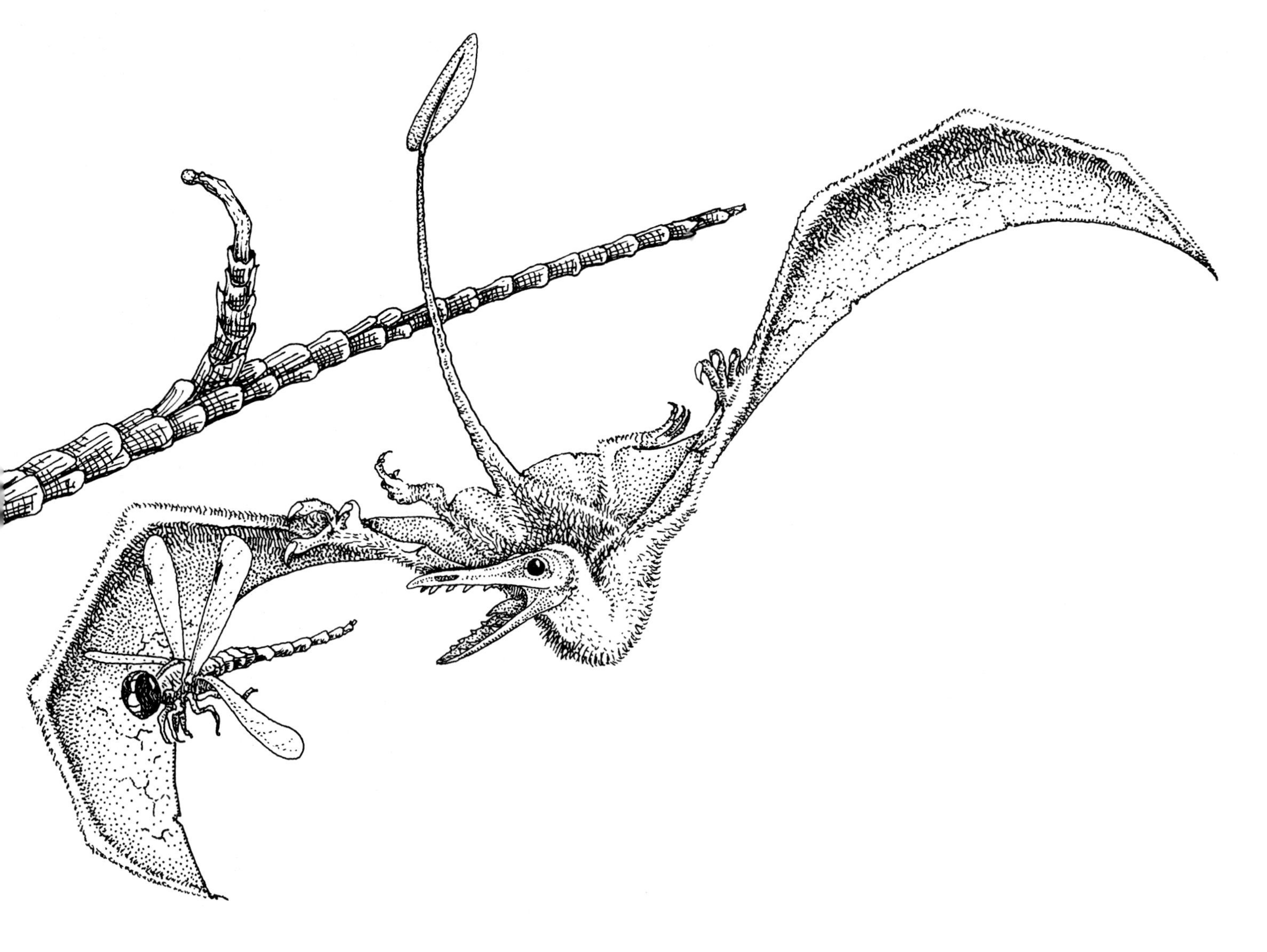

An early pterosaur, Rhamphorhynchus *("prow beak"), chases a dragonfly that is about the size of a modern dragonfly. The rise of the pterosaurs saw the fall of the giant dragonflies.*

For the first time, insects experienced selective pressure from other flying animals. The giant dragonflies and other biggies of the insect past, encased in their cumbersome armor and limited by a breathing system poorly suited to larger size, were unable to escape the agile pterosaurs with their light internal skeletons, lungs, and high energy. Almost immediately (on the evolutionary time scale) the average size of flying insects was reduced to what we see today.

As happens with all unchallenged groups, the pterosaurs underwent a great adaptive radiation in the next 150 million years, producing a variety of flyers: some that ate insects, some that ate fish, perhaps even some that ate plants. Well-preserved fossils of these creatures show us that they were covered with fur, like bats. This fur presumably protected them from loss of body heat. Its presence in the fossil record suggests the pterosaurs had achieved a degree of endothermy ("heating from within"), or internally controlled constant body temperature.

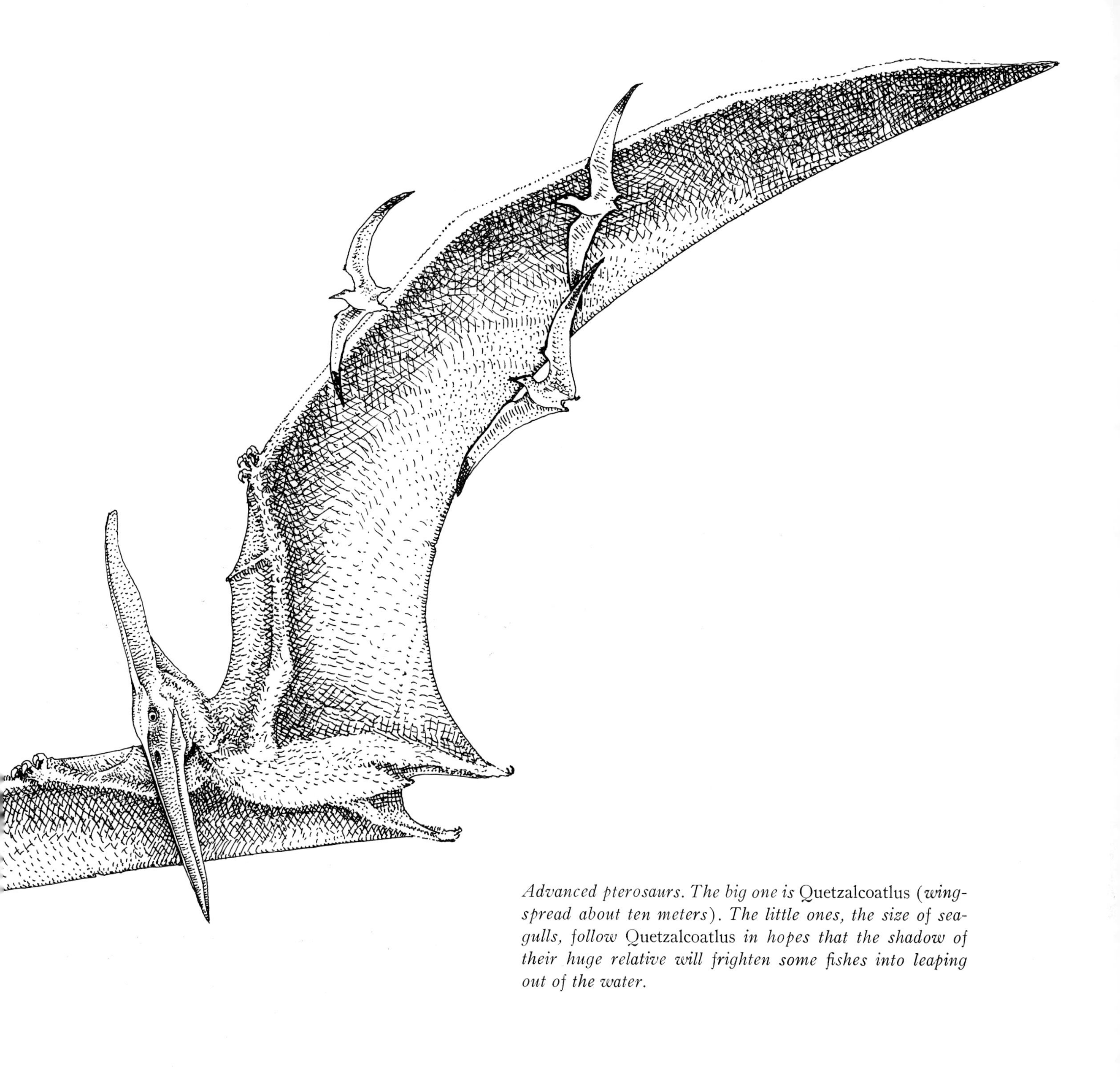

Advanced pterosaurs. The big one is Quetzalcoatlus *(wing-spread about ten meters). The little ones, the size of sea-gulls, follow* Quetzalcoatlus *in hopes that the shadow of their huge relative will frighten some fishes into leaping out of the water.*

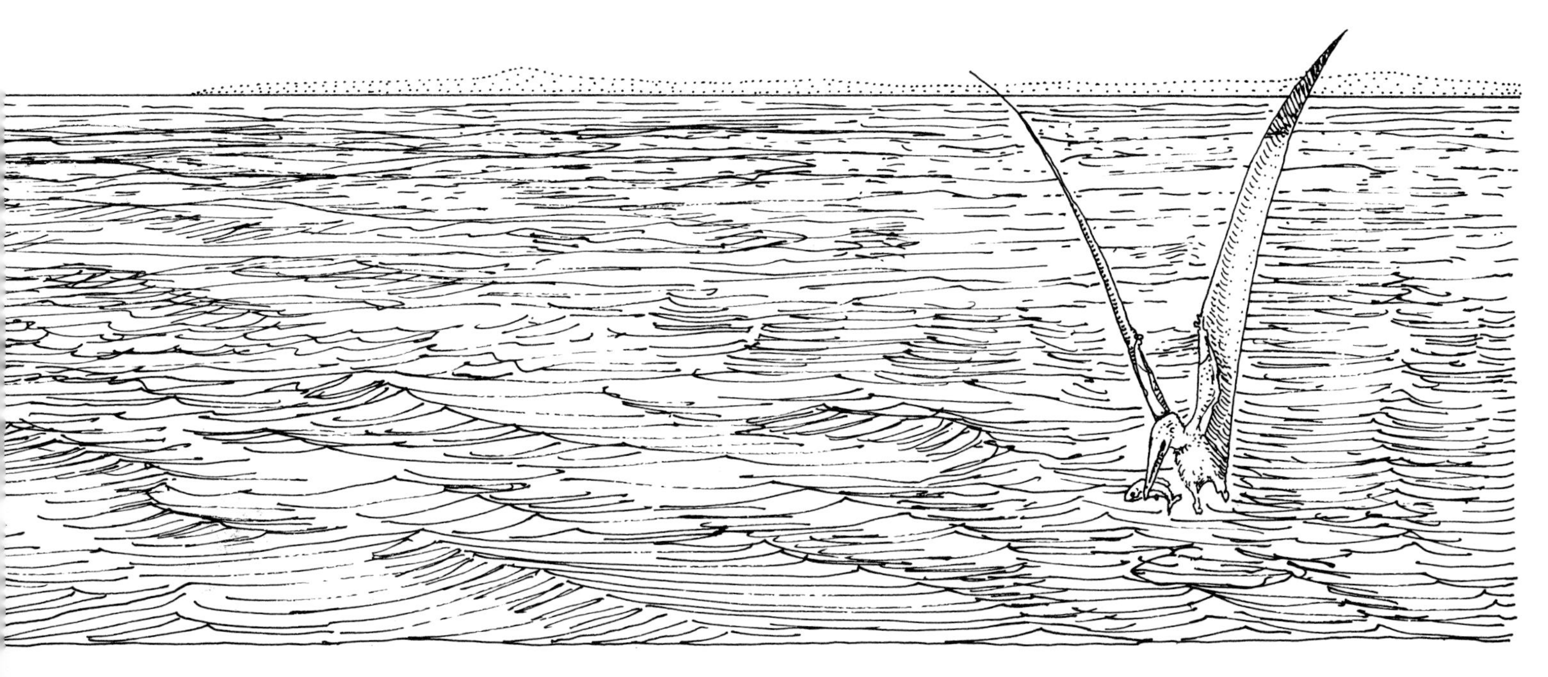

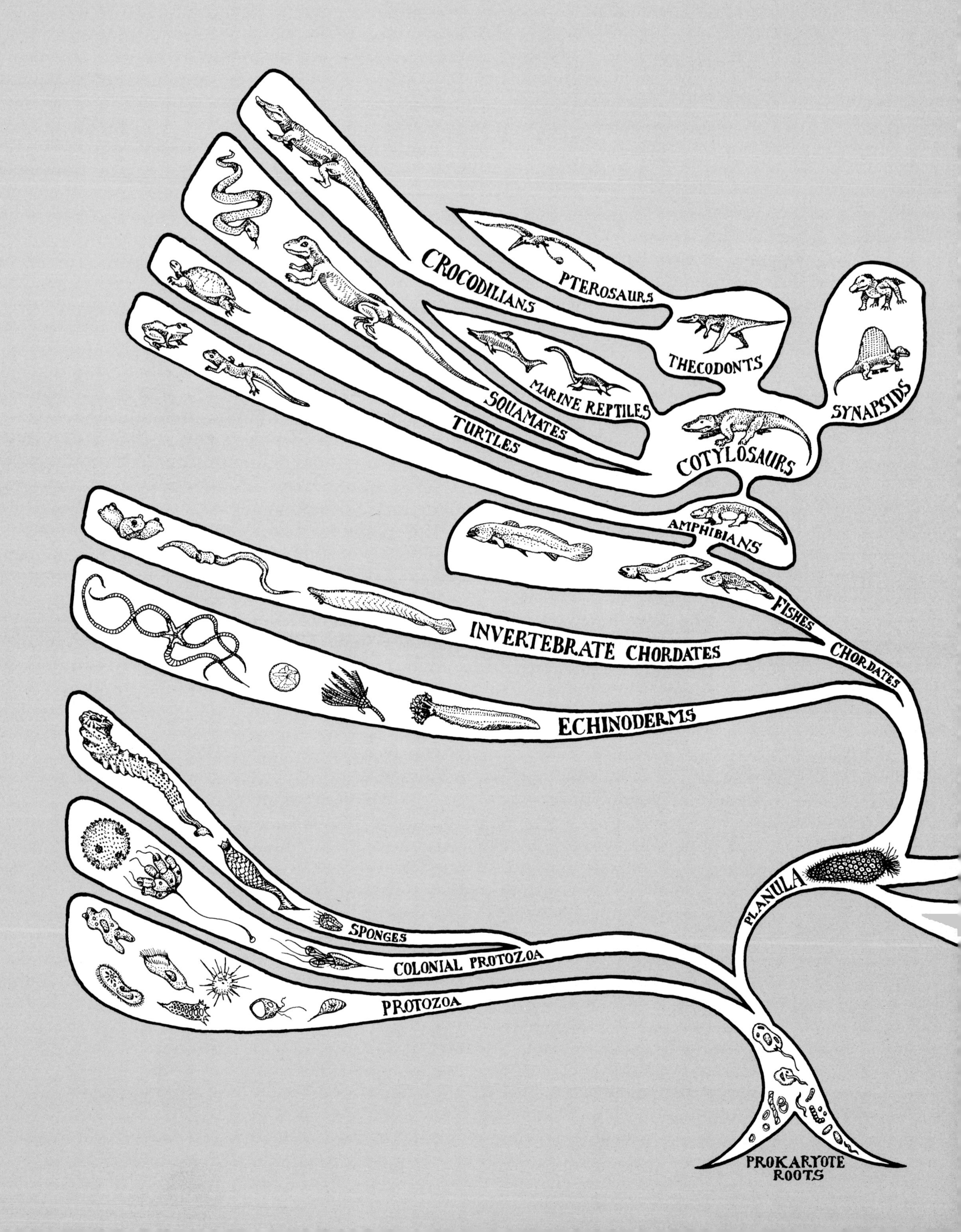
CROCODILIANS
PTEROSAURS
THECODONTS
SYNAPSIDS
SQUAMATES
MARINE REPTILES
TURTLES
COTYLOSAURS
AMPHIBIANS
FISHES
INVERTEBRATE CHORDATES
CHORDATES
ECHINODERMS
PLANULA
SPONGES
COLONIAL PROTOZOA
PROTOZOA
PROKARYOTE
ROOTS

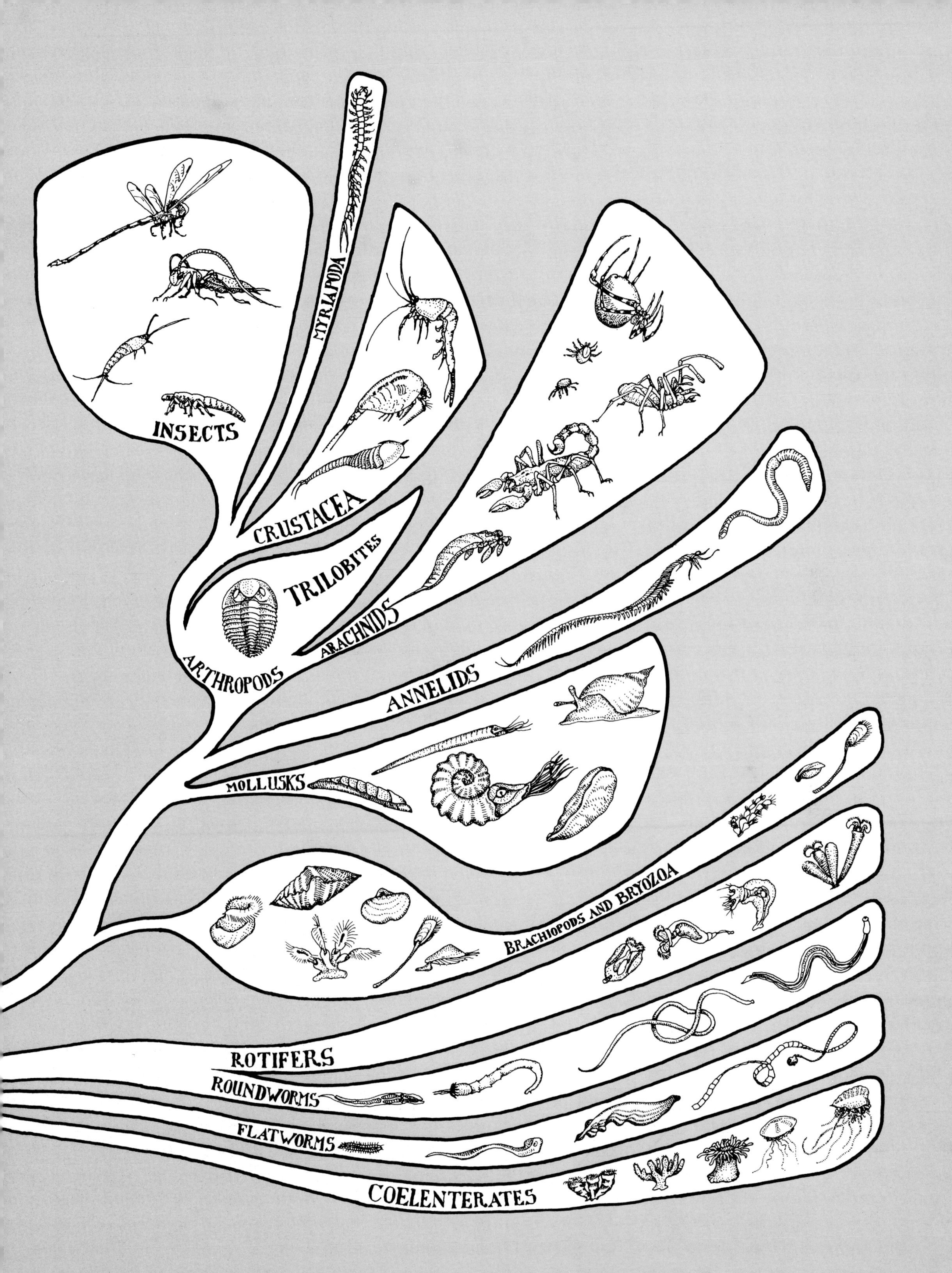
INSECTS
MYRIAPODA
CRUSTACEA
TRILOBITES
ARACHNIDS
ARTHROPODS
ANNELIDS
MOLLUSKS
BRACHIOPODS AND BRYOZOA
ROTIFERS
ROUNDWORMS
FLATWORMS
COELENTERATES

Meanwhile, the ground-dwelling thecodonts continued to improve their fast hunting way of life. They perfected a two-legged walk and no longer reverted to four legs when moving slowly. As a result, their forelegs became permanently available as insect-catching "hands," and they became the first full-time bipedal runners among the vertebrates. The long thecodont tail, no longer used as a swimming oar, stiffened into a balance rod that countered the weight of the head and forelegs.

As the evolution of the synapsids showed, high-energy living requires some method of breaking food into small pieces for rapid digestion. Synapsids accomplished this by evolving heavy heads with large muscles and grinding molars. Thecodonts, however, were forced by their two-legged posture to maintain a light skull framework. Their heaviest body parts needed to be near the center of gravity at their hips. Still, their high-energy running required some

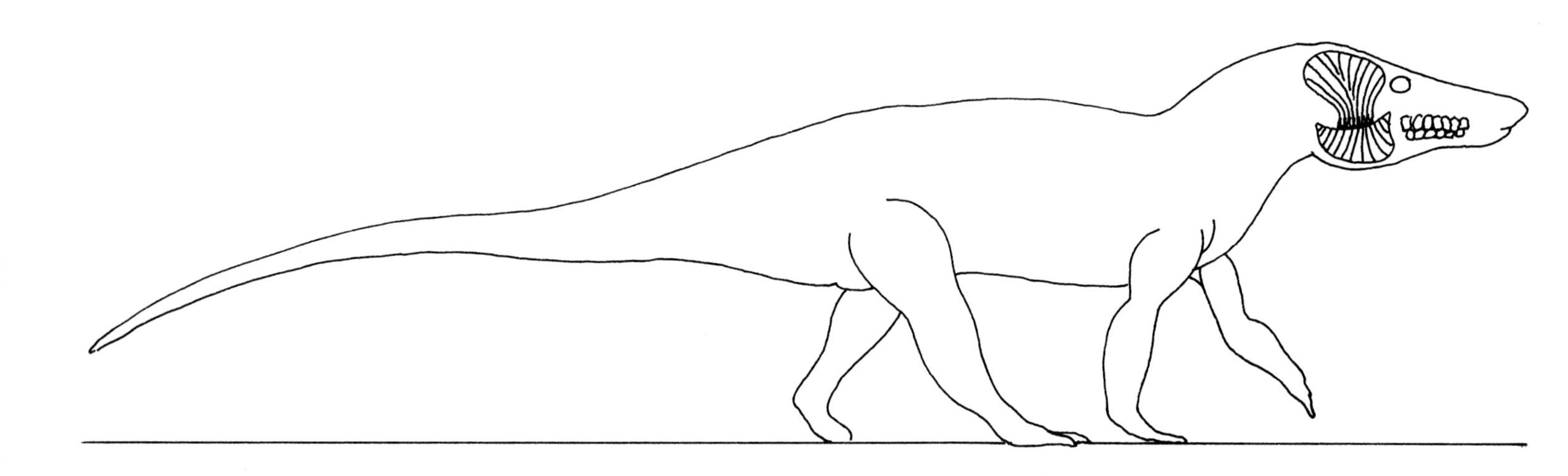

The mechanisms by which synapsids (left) *and thecodonts* (right) *broke down their food. The synapsids chewed with heavy teeth and jaw muscles. The thecodonts required a light skull because of their bipedal gait. They evolved a gizzard* (cross-section, right), *within which stones ground the food to a pulp. Because the gizzard was close to the thecodont's center of gravity, the animal could run far more easily than it could have with a headfull of weighty teeth and massive jaw muscles. In the thecodont, the head was an organ for seizing and killing, not for chewing.*

method of chewing food. Mutations occurred that created a special digestive chamber—the gizzard—located within the rib-cage near the body's center of mass. Here, food was ground up by powerful muscles in a sort of mill, aided by stones that the animals swallowed for this specific purpose. These stones are often found in the fossils of thecodonts and their descendants, and are called gastroliths ("belly stones"). They mark the existence in these animals of that high-energy lifestyle that was to take them to dominance of the vertebrate world. Birds, remote descendants of thecodonts, show us the gizzard in action. Birds, indeed, show us a lot about archosaurian ways, so much so that some scientists regard birds as flying archosaurs, much as bats are flying mammals.

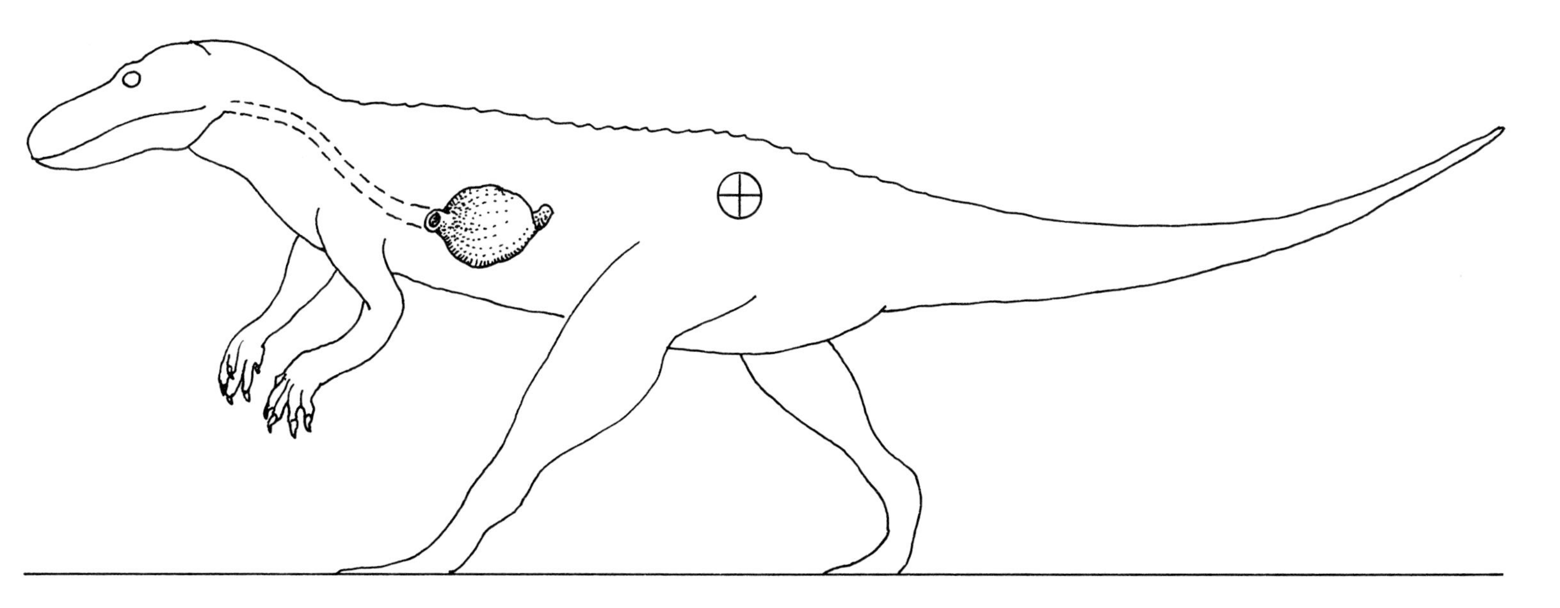

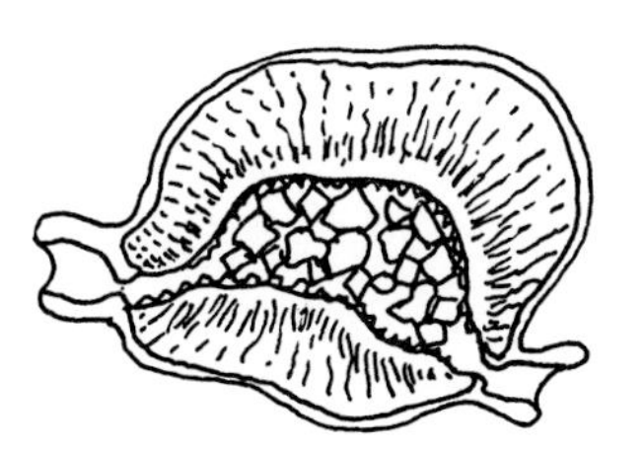

Early dinosaurs of the genus Coelophysis, *("hollow bones"). These predators inhabited the western North American continent about 200 million years ago. Although they were the size of dogs, they probably weighed considerably less, being extremely speedy and*

By about 200 million years ago, we find an assemblage of little, sharp-eyed running archosaurs, many about the size of chickens—energetic, always hungry. These were the first dinosaurs, the finest, fastest, smoothest creatures the world had ever seen. Oh, they were good at the game of life!

So excellent were they, in fact, that they quickly came into competition with one another and with the therapsids that up until now had been so successful. Being faster, and possessing taloned "hands" as well as sharp teeth for killing prey (not for chewing), dinosaurs gradually squeezed the therapsids into extinction and inherited their econiches.

birdlike in build. Dinosaurs are traditionally considered "reptiles," but the contrast between these animals and their more reptilian prey in this picture amply points up the problem with that classification.

An early mammal, only a few centimeters long, hides with her young among the tree roots of a primordial forest. A cat-sized predatory dinosaur peers in at the little family.

Although the therapsid triumph was over by 200 million years ago, a few of their kind had managed to evade the dinosaurs by invading a new econiche—the night, with its cool temperatures and darkness. These therapsid descendants were for the most part small insect eaters; their size permitted some of them to escape the notice of night-hunting dinosaurs that located their prey by means of highly developed eyes and ears. As these little therapsid descendants evolved for nighttime activity, those possessing some sort of body insulation fared better than those that had none. Eventually fur appeared, trapping warm air close to the skin. Gradually the dinosaur oppression of the last of the therapsids produced the first mammals, our own ancestors, which began as furry, shrewlike creatures of the night.

The night exerted other pressures on these animals. Because they lived in the dark, their weak eyes were of little use to them. Instead, they relied on their senses of hearing and smell, and these became sharper and keener. Through 140 million years of dinosaur dominion, the mammalian forebrain, the part associated with smell, was en-

The sharp eyes and talons of such dinosaurs kept the early mammals in the night for some 140 million years. As a result, their nocturnal senses developed and their brains enlarged.

larged and enhanced in a world that allowed of these nearly blind creatures absolutely no mistakes.

Again, the terrible night killed many early mammalian young, which were born alive and unprotected. In response to this selective pressure, infant care evolved, and also, somewhere along the line, the feeding of the young with milk from the mother's body. These adaptations may even have arisen back in therapsid times. However, the rule of the dinosaurs certainly increased the dependence of those young on their mothers. In addition to relying on instinct to get about, young mammals learned more and more from their parents. Because their main sense organ was the nose, and all information was received through it, learning took place largely in the smelling centers of their brains.

For the next 140 or so million years, these mammals remained tiny, mostly shrew-sized, none larger than a cat. They were marginal, unimportant beings in the dinosaur world, and they kept to the forests of the night.

Meanwhile, the dinosaurs had begun to diversify as their growing populations brought them into competition with one another. Mutations appeared that permitted some dinosaurs to eat plants, and this new capacity proved valuable. Plant-eating dinosaurs did not need to compete with their carnivorous relatives for food. Their plant-eating econiche gave them a new means of survival.

At the time of their conquest around 200 million years ago, dinosaurs lived in a world rich in plant life. Modern conifers (evergreen trees like pines and spruces) and flowering plants were on the rise. Herbivorous (plant-eating) dinosaurs rose with them. Some took to eating the conifers. This dependence on a tall-tree diet favored forms with long necks and great size, or mutations that led to those forms. Other herbivorous dinosaurs became adapted to the many new flowering plants that were appearing.

Two different stocks of primitive vegetarian dinosaurs. (this page) *The long-necked* Plateosaurus, *likely a browser of tall trees, walked on its hind legs like its carnivorous ancestors, but also spent some time on four legs. From it descended the well-known sauropods, the long-necked dinosaurs epitomized by the famous brontosaurs.* (right) *Several* Hypsilophodons, *dog-sized browsers with nipping beaks in front and shearing tooth rows for cutting coarse plant matter. They may have lived like deer.*

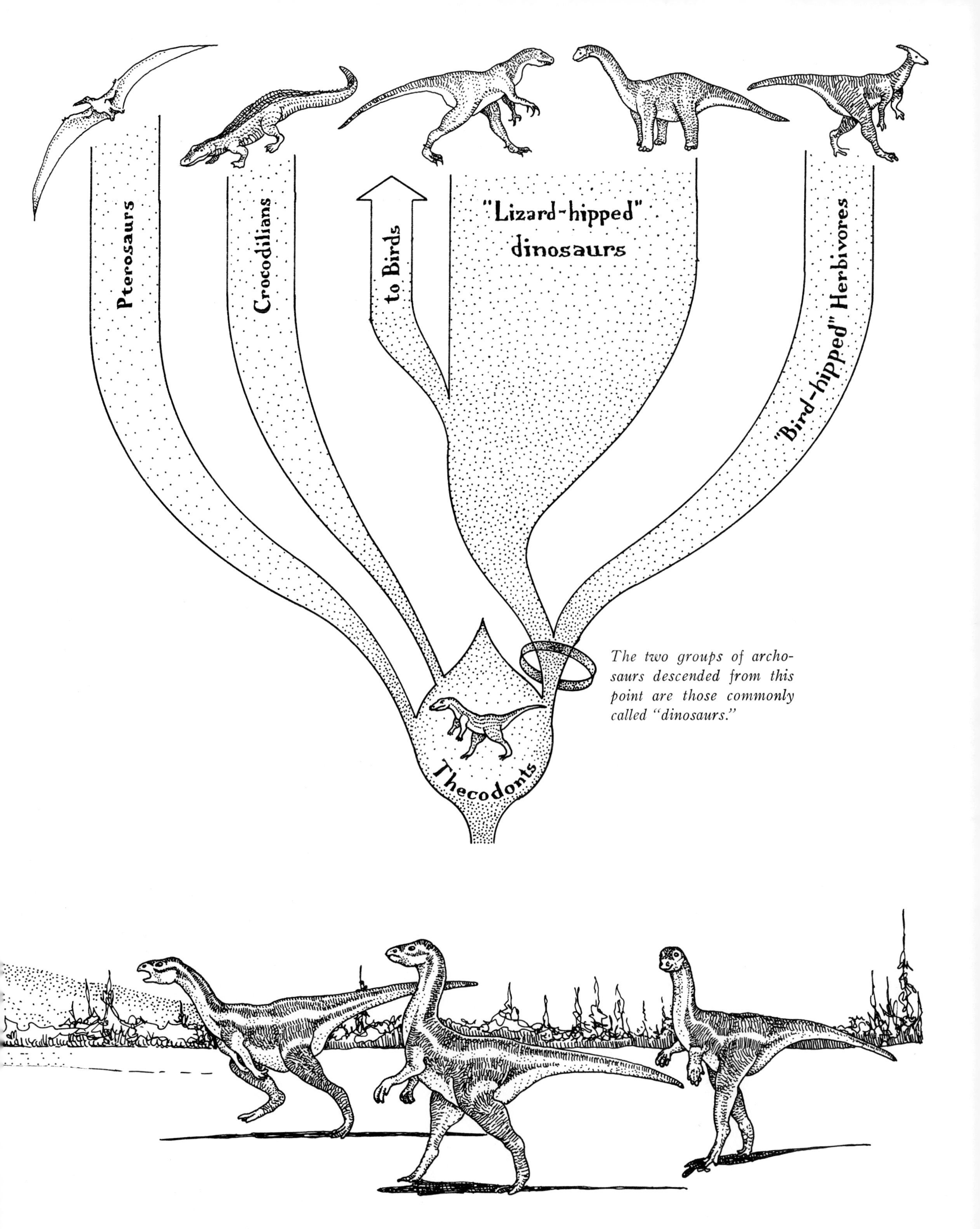

(above) *Chart depicting the archosaurian adaptive radiation from its thecodont roots to world dominion.*

By about 100 million years ago, the world's continents (old Pangaea was breaking up into the modern landmasses by then) were populated by grazing dinosaurs, browsing dinosaurs, dinosaurs that ate thorny plants and soft plants, and dinosaurs that, it appears from their remains, even ate wood! Many of these herbivores moved about in great herds, the bulls on the outskirts to protect the smaller females and young from the packs of ever-hungry predators that followed them.

A large browser-grazer of genus Parasaurolophus *("sort-of-a-lizard-with-a-crest"), an advanced herbivorous dinosaur of about 70 million years ago, guards her eggs. In the background, a* Parasaurolophus *herd traverses a savanna. The fossils of such dinosaurs are sometimes found with their eggs and young, implying that they practiced a high form of parental care.*

And such predators those were! From the earliest chicken-sized, insect-hunting dinosaurs had arisen carnivores of all sizes, some as small as modern pigeons, some nearly as big as elephants. The biggest predators, like the famous tyrannosaurs, probably ate dead meat as well as live animals, being too slow to chase swift prey very well. Most hunting dinosaurs were lion-sized or smaller and seem to have been fast runners. It is likely that many of them stalked their prey in groups, so as to be able to kill the biggest herbivorous dinosaurs. Some seem to have entered the night, perhaps specializing in killing and eating our own ancestors, the little mammals. These dinosaurs were small, with large eyes and sharp talons like those of their daytime counterparts.

A pack of hunters, Deinocheirus *("terrible hand," so named because of the frightful talons on their forefeet), have just slain a large herbivore and prepare to stand off some competitors (out of the picture). These carnivores stood three meters high and were*

swift and agile. Attacking their prey on the run, they seem to have killed by grasping with their "hands" and kicking with the large recurved talons of their hindfeet, folding the two smaller toes out of the way in the process.

Archaeopteryx, *"old wing," was either an early bird or a small dinosaur, depending on how you look at it. In this drawing, it is attacked by a larger relative, itself a very birdlike creature with regard to posture and general demeanor.*

From small predatory dinosaurs, too, appear to have arisen the first birds. Fossils of the first birds are remarkably similar to fossils of dinosaurs. In fact, were it not for the fine lithographic shales that preserved the imprint of feathers, the first birds would have been classified as small dinosaurs. Actually, bird feathers appear to have originated as insulating scales, which may also have been present on most small dinosaurs. Bird wings evolved from the "hands" of small carnivorous dinosaurs. Their toothless beaks arose because teeth (which are heavy) were selected against as early birds perfected flight. Today some scientists include birds within an order Dinosauria, in a class Archosauria, emphasizing that, except for flight-related specializations, birds are little different from their dinosaurian relatives.

(top) Ichthyornis (*"fishbird"*), *a toothed ternlike seabird of 80 or so million years ago.*
(above) Hesperornis (*"western bird"*), *a large, toothed, aquatic bird of the same period.*

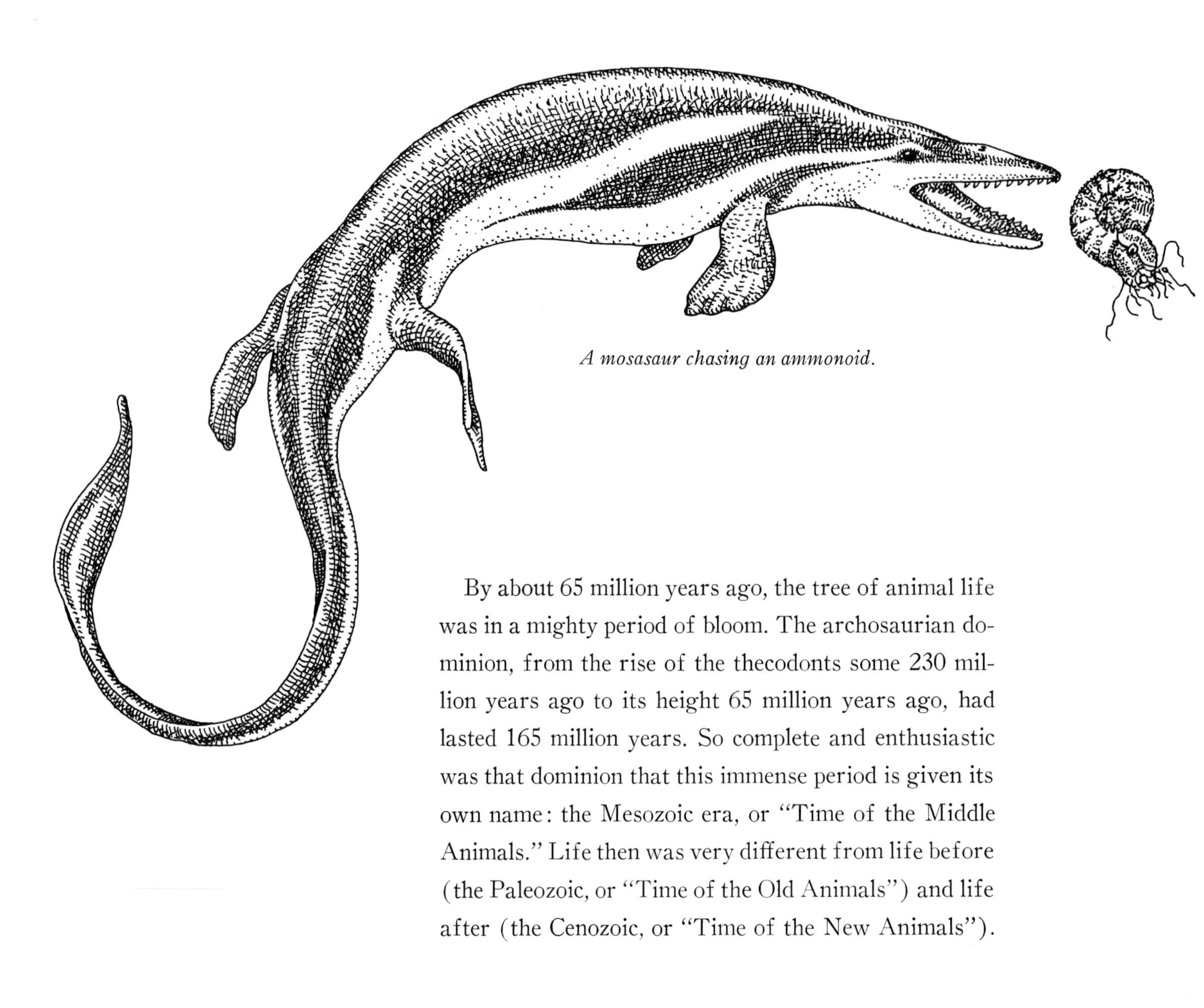

A mosasaur chasing an ammonoid.

By about 65 million years ago, the tree of animal life was in a mighty period of bloom. The archosaurian dominion, from the rise of the thecodonts some 230 million years ago to its height 65 million years ago, had lasted 165 million years. So complete and enthusiastic was that dominion that this immense period is given its own name: the Mesozoic era, or "Time of the Middle Animals." Life then was very different from life before (the Paleozoic, or "Time of the Old Animals") and life after (the Cenozoic, or "Time of the New Animals").

SOME OF THE LARGER INHABITANTS OF THE EARTH'S SEAS DURING THE MESOZOIC.

An ichthyosaur

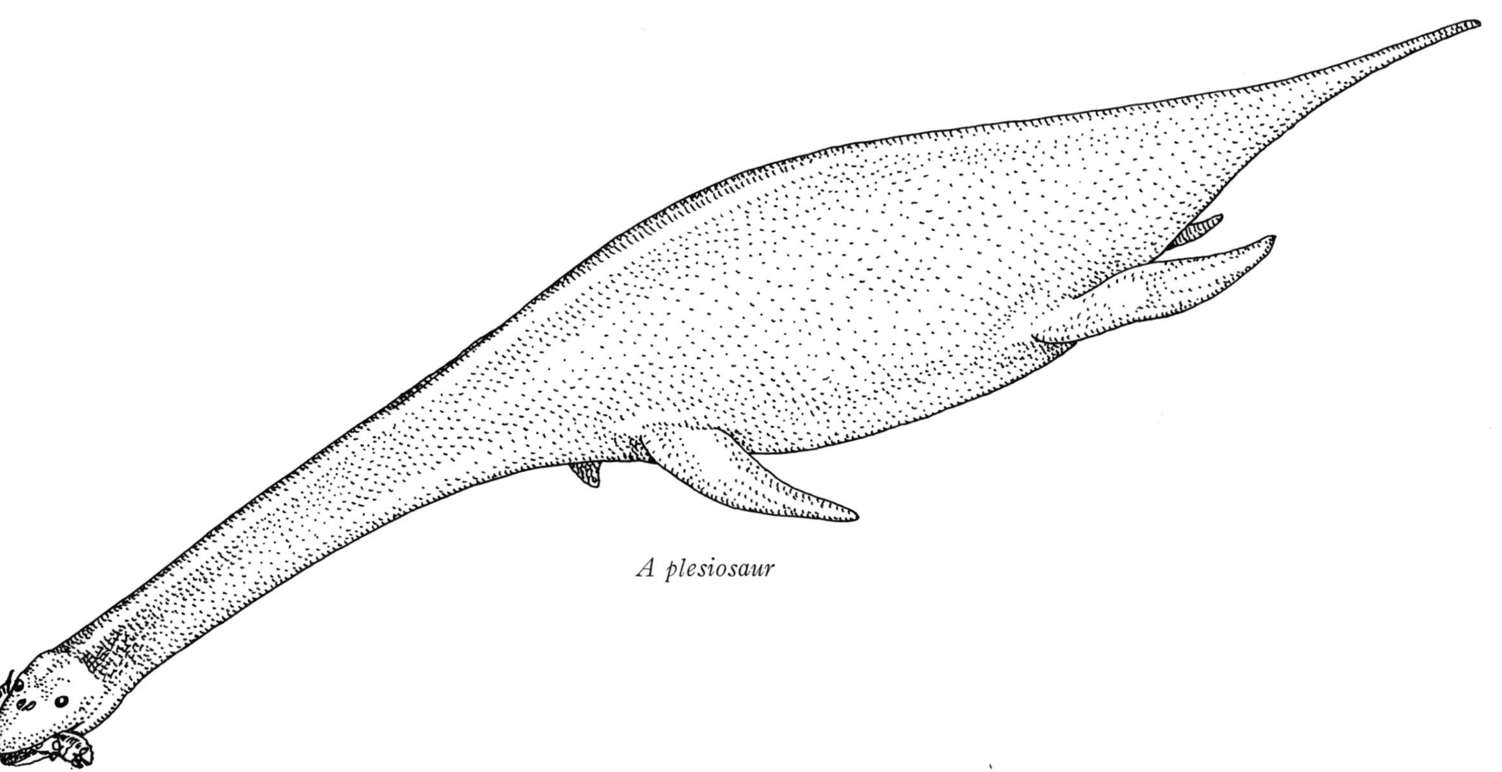
A plesiosaur

On land, the Mesozoic was a time of dinosaurs, and of conifer trees and flowering plants much like those of today. In the air, it was a time of birds. In the seas, it was a time of diversity, with the modern sharks coming into sway as superb predators, and with thousands of varieties of coiled molluscan ammonoids feeding on essentially modern fishes and modern crustaceans. Large marine reptiles, occupying the econiches that are today occupied by modern air-breathing forms such as seals and whales, anticipated the size and streamlined shape of these forms (an example of convergence). The oceans teemed with seal-like plesiosaurs ("near-lizards") and whale-like mosasaurs. For a while there were many varieties of dolphin-like ichthyosaurs, "fish-lizards." All of these creatures were descendents of the cotylosaurs that had taken to water so long ago.

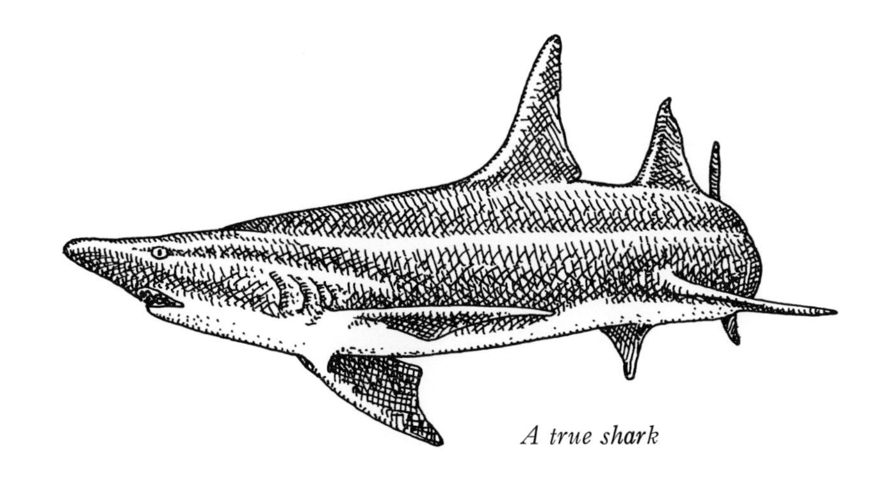
A true shark

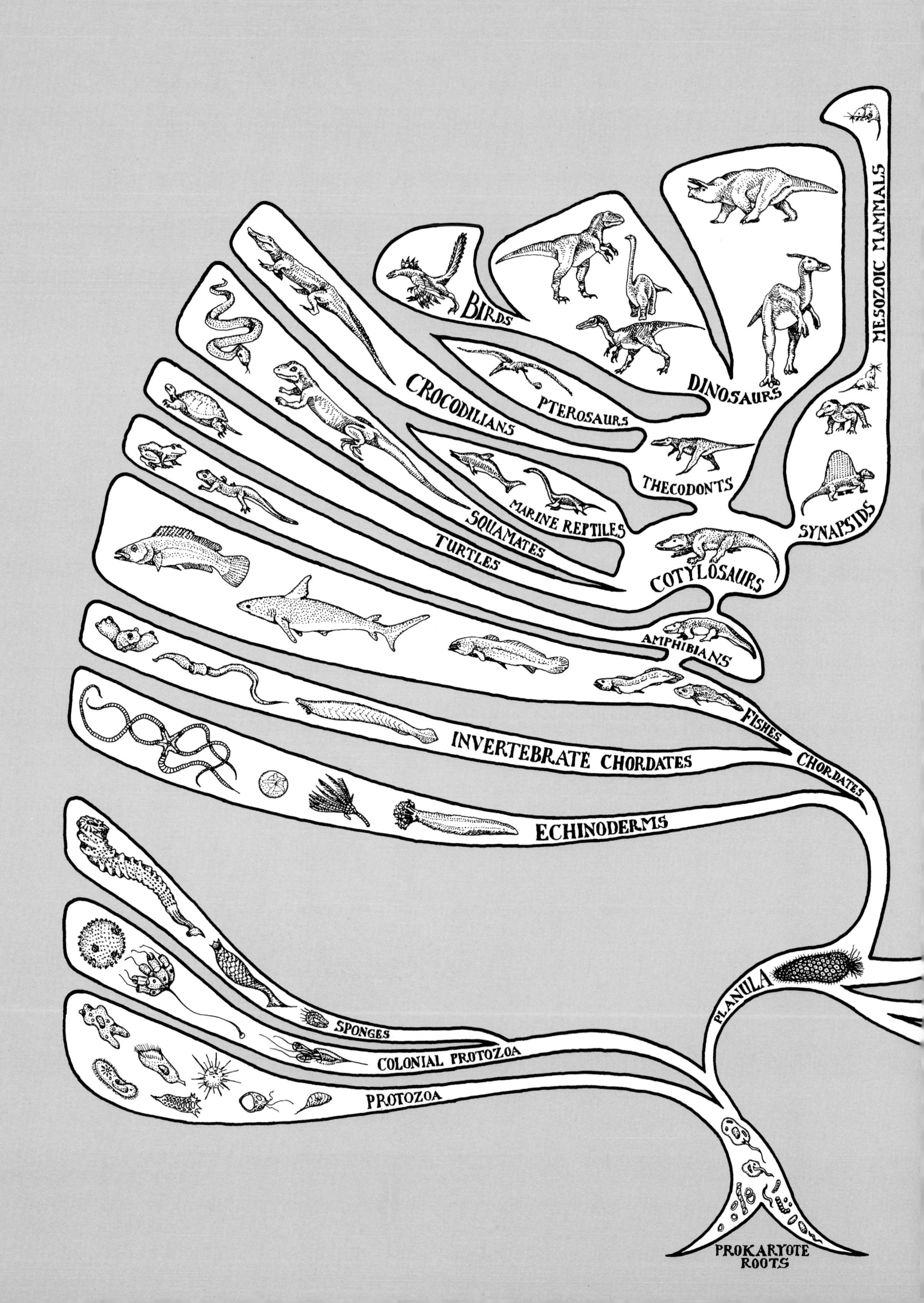

MESOZOIC MAMMALS
BIRDS
DINOSAURS
CROCODILIANS
PTEROSAURS
THECODONTS
MARINE REPTILES
SQUAMATES
TURTLES
SYNAPSIDS
COTYLOSAURS
AMPHIBIANS
FISHES
INVERTEBRATE CHORDATES
CHORDATES
ECHINODERMS
PLANULA
SPONGES
COLONIAL PROTOZOA
PROTOZOA
PROKARYOTE
ROOTS

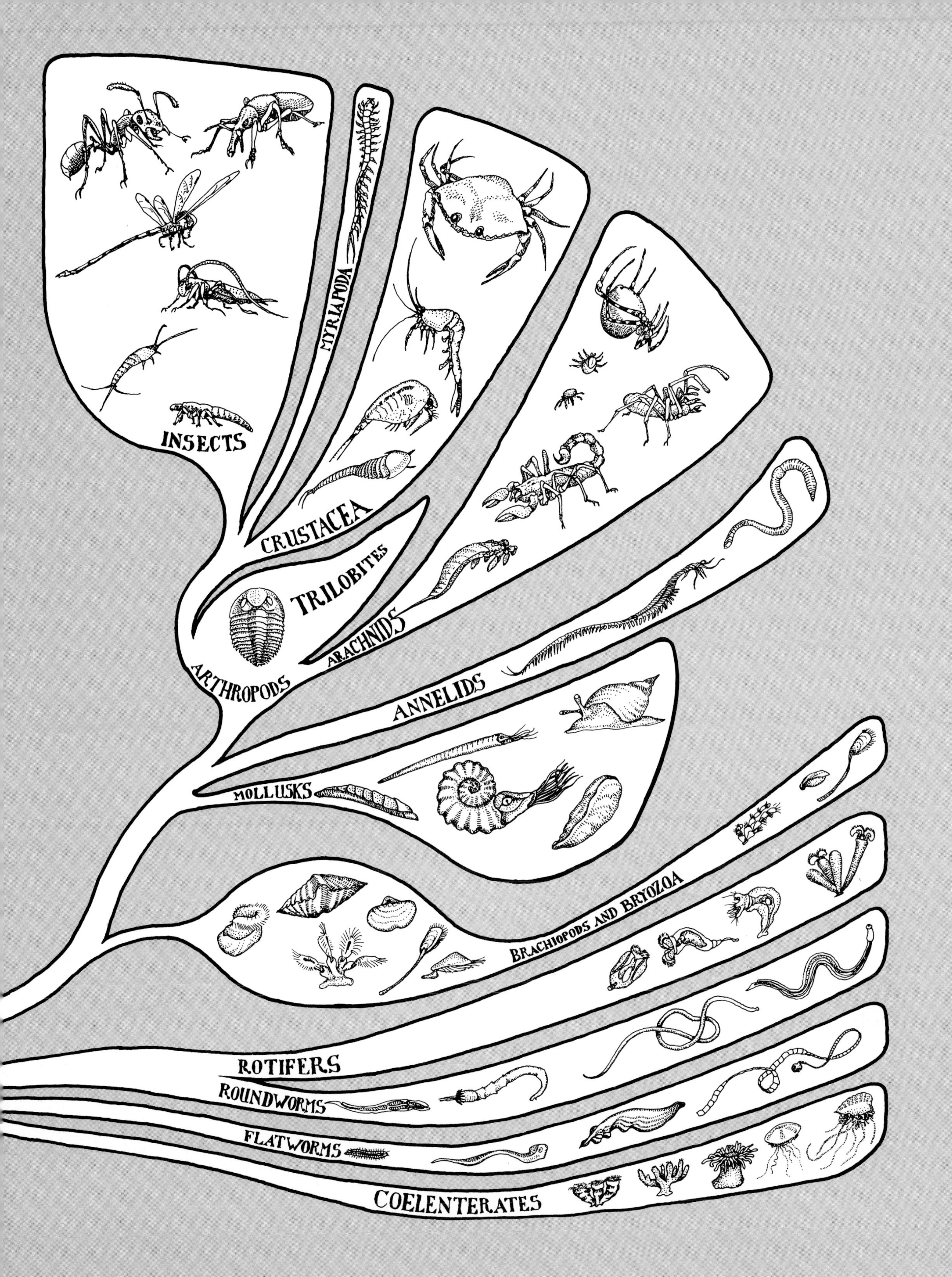

INSECTS
MYRIAPODA
CRUSTACEA
TRILOBITES
ARACHNIDS
ARTHROPODS
ANNELIDS
MOLLUSKS
BRACHIOPODS AND BRYOZOA
ROTIFERS
ROUNDWORMS
FLATWORMS
COELENTERATES

If the Mesozoic arose with the archosaurs about 230 million years ago, when did it end? And why? The Mesozoic ended in death, a catastrophic laying waste of the earth's surface 65 million years ago. Evidence is rapidly mounting to show that an asteroid perhaps ten kilometers in diameter may have struck one of the earth's oceans. When an asteroid strikes, its energy is changed to heat, heat which could evaporate the ocean around it, producing a cloud cover. This cloud cover would have restricted the sun's impact on plants, impeding photosynthesis and starving out all the hungriest, most energetic animals. Those forms that must eat the most to survive are always the first to be affected when something interrupts the food chain. The dinosaurs and the great seagoing reptiles became extinct, as did the predatory ammonoids.

The end of the Mesozoic. Among the last dinosaurs were those of genus Triceratops *("three-horn face"), large herbivores that seem from fossils of their teeth to have been able to eat plant substances as tough as wood. Here, a starving specimen staggers past the corpses of some of its relatives. A small scavenging dinosaur feeds on a carcass, but will itself soon run out of food. In left corner, a little mammal watches from the head*

Small animals like insects, crustaceans, birds, and mammals, on the other hand, were able to live on the remaining scant supplies of food until the cloud cover broke and photosynthesis was restored. Reptiles that could survive for a time without food, such as crocodiles, snakes, lizards, and turtles, made it through the end of the Mesozoic in fine style. Although plants ceased photosynthesizing for a while, their roots, seeds, and other parts were equipped for periods of dormancy. These reawoke when the sunlight came back.

of another dead Triceratops; *in time, its descendants will fill the ecologic gap left by the dinosaurs.* Triceratops *and many of its relatives sported a bony "frill" on the skull that paleontologists originally thought was defensive in nature, protecting the animal's neck. Actually, the frill was an anchor for the mighty muscles that operated its jaws and supported its massive head. The horns on its face were more than sufficient for protection. Like modern bison, these horned dinosaurs traveled in vast herds—until the end.*

A Diatryma, *or "terror crane," runs down a small Paleocene mammal. With a head as large as that of a horse, the* Diatryma *was an unpleasant character indeed. Luckily, the mammals edged such birds out of competition early in the Cenozoic.*

Whatever happened to end the Mesozoic, the dawn of the Cenozoic, or "Time of the New Animals," saw the world a lonely place. The tree of animal life had been sharply pruned back by the asteroid (or whatever it was). The Cenozoic opened still rich in plant life, well equipped with insects, reptiles, birds, and little mammals. But there were no large, fast animals left anywhere. On the land and in the seas, their econiches were all empty, waiting.

The high-energy land animals were now birds and the little mammals. Both groups made a bid for the old dinosaur econiches. For a time, gigantic terrible flightless birds roamed the world in the epoch known as the Paleocene, or "Old Dawn of the Recent," the period of about ten million years after the great death.

Because the mammals had teeth and were therefore more efficient than the toothless ground-birds, they eventually won out. By the late Mesozoic several modern groups of mammals had already appeared, and these quickly diversified during the Paleocene. Carnivorous mammals arose, as well as specialized seed-eating mammals (the squirrel-like earliest rodents), and flying mammals (bats), and running herbivorous mammals (ungulates, or hoofed mammals like horses and deer). Over time, their predators evolved and got good at catching and eating them; and they, in turn, evolved to escape those predators.

A scene in western North America around 10 million years ago. A "bear dog" (left), *a large slow-moving predator, watches as a saber-toothed cat attempts to drive a young chalicothere ("clawed beast," a distant relative of modern horses) from the safety of*

its mother. In the background, a pair of Synthetoceras, *related to modern deer and cattle, look on warily.*

By about 12 million years ago, all the Mesozoic econiches were filled again. There were not only mammals occupying the old dinosaur niches on land, there were mammals in the seas occupying the old reptilian niches in the form of the modern seals and whales. The tree of animal life had compensated for its terrible pruning at the end of the Mesozoic, and life went on joyously for the remainder of the Cenozoic era, until about four million years ago . . . when something new happened to the tree.

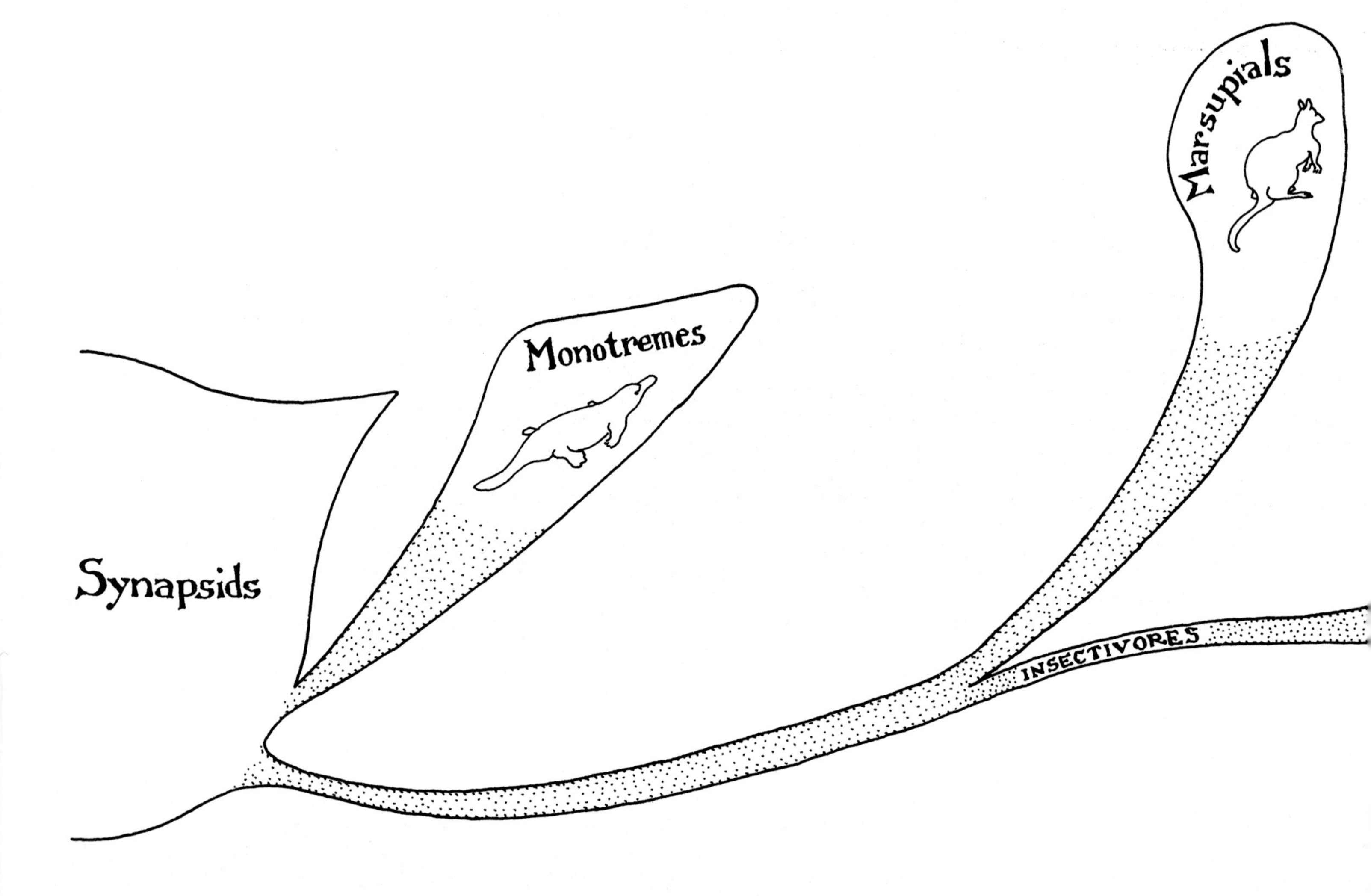

THE MAMMALIAN ADAPTIVE RADIATION.

(For ease of recognition modern forms are shown in each branch.)

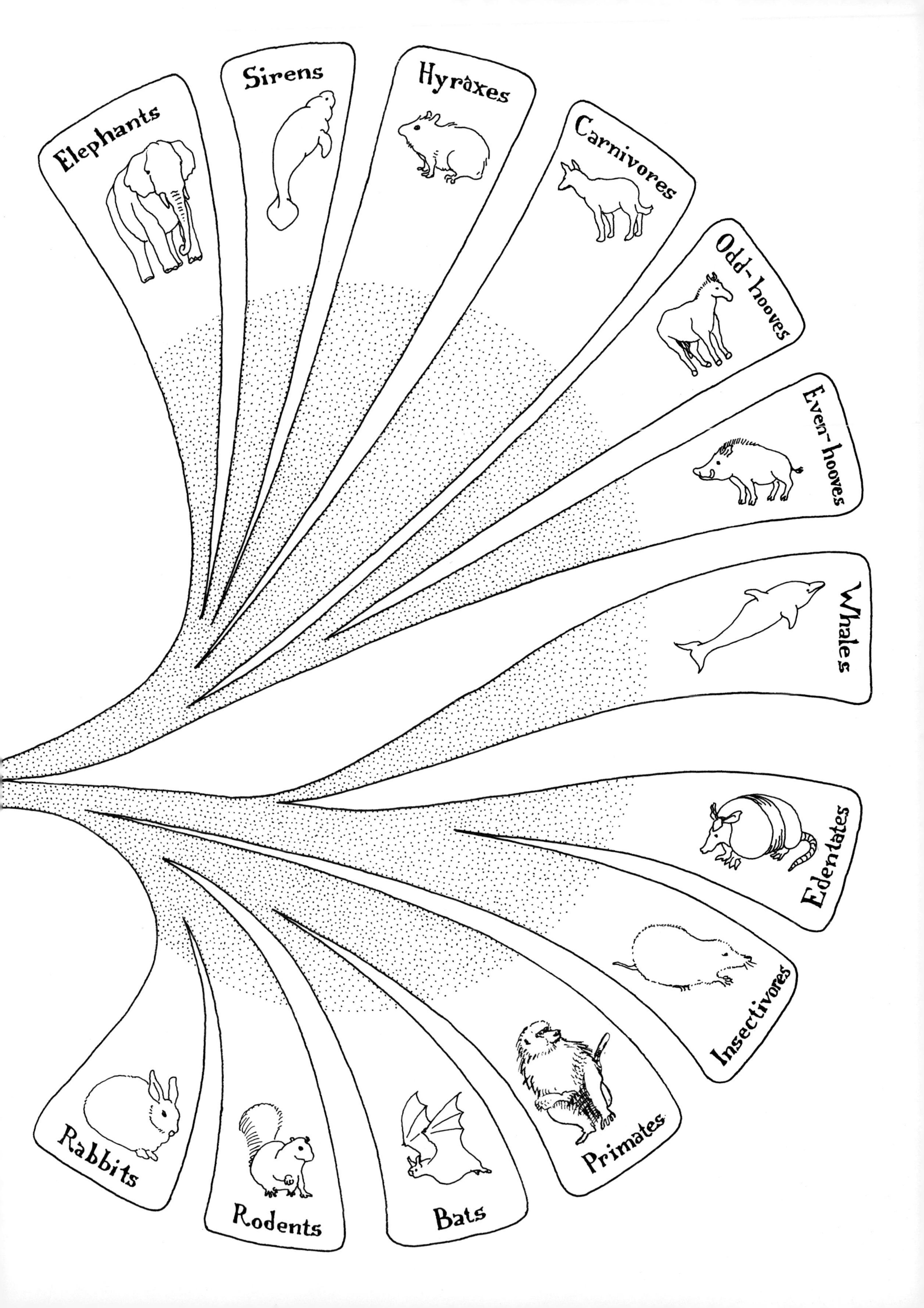
Elephants
Sirens
Hyraxes
Carnivores
Odd-hooves
Even-hooves
Whales
Edentates
Insectivores
Primates
Bats
Rodents
Rabbits

A generalized squirrel-like tree-dwelling primate of the late Mesozoic.

To understand this completely novel event, we return to the last ten million years or so of the Mesozoic. Here we look at the mammals of the time: small, for the most part nocturnal dwellers on the forest floor. Some, however, had taken to climbing trees. In trees there were no dinosaurs. Instead, there were birds' eggs, insects, fruits, lizards, snakes, and other edibles, making the arboreal econiche a favorable one for these little creatures.

Tree-dwelling, however, exerts strong selective pressure on its practitioners. Either you fall and die, or you possess skills that allow you to survive in the treetops long enough to bear young that also possess those skills. The mammalian generations that pioneered tree-dwelling were rigorously selected for balance, good eyesight, good grip, and for the ability to guess in advance the trajectories of their leaps and swings from branch to branch. By the end of the Mesozoic, this selection had resulted in mammals with large, forward-directed eyes, which gave them two overlapping fields of vision so that they might calculate depth as they moved about high above the ground. To spot edibles in their leafy environment, they possessed acute color vision, a rarity among their smell-dependent mammalian kind. For tight, sensitive gripping of the tree limbs upon which their lives depended, these agile beings came equipped with padded fingers and toes, one of which on each foot was opposed to the other. With them, branches could be encircled to lessen the chances of falling.

Because this is one of the first mammalian groups of modern kind to have appeared, and because the group includes the most visibly intelligent of animals on earth, these tree-dwellers are called Primates ("first") by scientists.

A primitive lemur, Notharctus, *showing the attributes of most primates—forward-directed eyes and grasping hands for maneuverability in the treetops.*

For much of the Cenozoic, primates spent all their time in the trees, evolving into a variety of species dependent on different types of arboreal habitats, different food supplies, different climates. All of them, however, continued to experience the unique selective pressures of tree-dwelling. And all of them, to a greater or lesser extent, responded by continuing to evolve in the special primate direction of sensitive hands and feet, keen eyes and foot-to-eye coordination, large brain, and omnivorous ("eats anything") diet.

Some of these primates remained small and comparatively primitive, keeping to the night like their Mesozoic ancestors. Among them are the tarsiers and lemurs, the "lower primates."

The rise of the primates. (left to right) *A tarsier, one of the most primitive of living primates; a ruffled lemur; a sifaka; a colobus monkey; a gibbon, one of the more primitive anthropoid (manlike) apes; and a chimpanzee, our closest living nonhuman relative.*

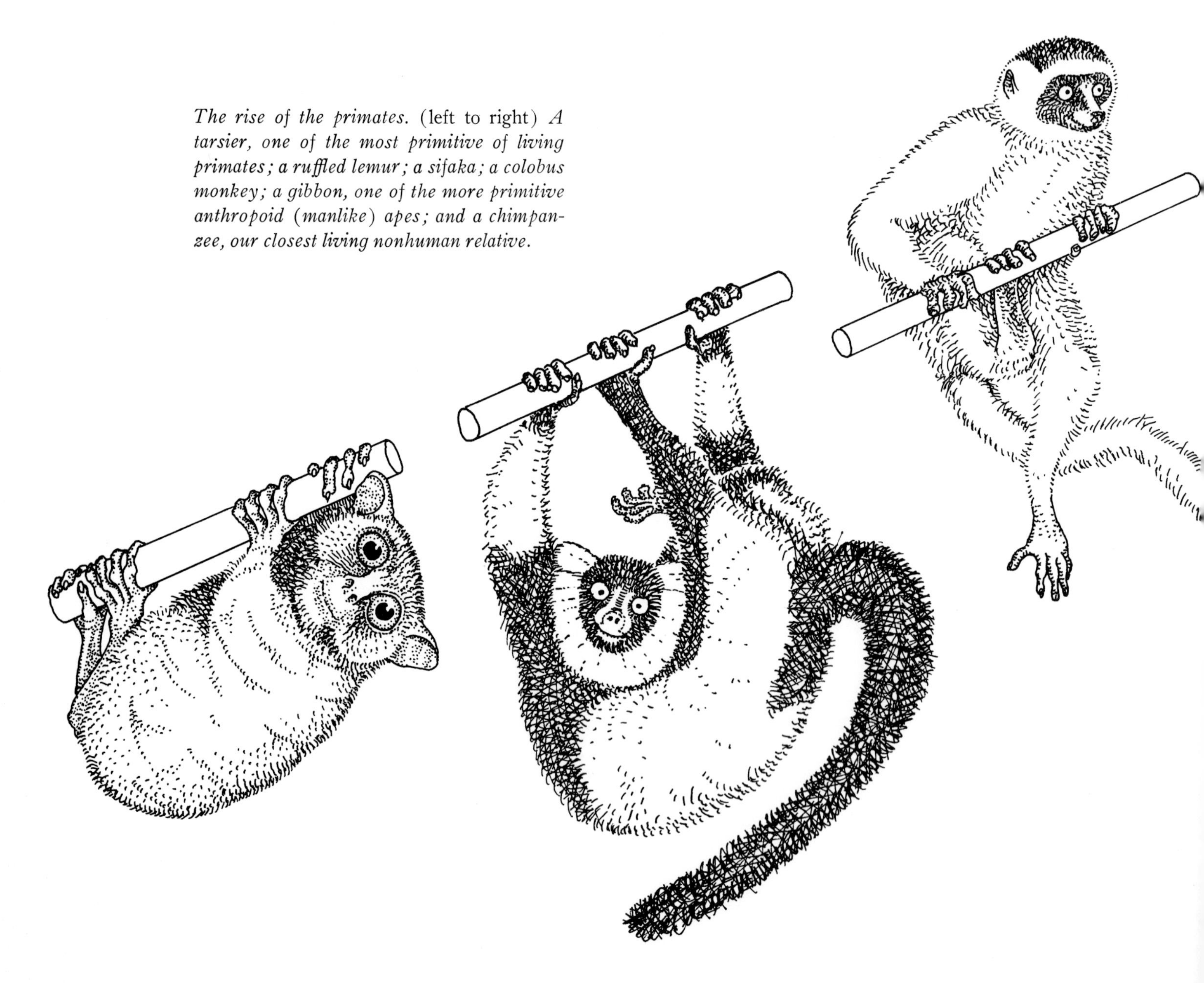

Others, however, moved into the daylight, keeping track of one another by hooting and gesticulating, forming family societies in the treetops. From primates like these evolved our modern monkeys: jolly talkers in the trees, agile, sharpsighted, and intelligent. From them, too, evolved larger forms, the apes: chimpanzees, gorillas, orangutans, and gibbons, whose inventiveness is well known. And also from them evolved, perhaps ten or so million years ago, primates that left the trees, entering the grasslands across which roamed vast herds of grazing mammals.

An African savanna scene around 5 million years ago. A troop of baboons (left), *wary but unafraid, observes a*

Most people have seen baboons in pictures or at zoos. Like other monkeys, baboons are social and intelligent. But unlike other monkeys, baboons spend a lot of time on the ground, on grassy plains, rooting about for food and yelling at one another. Baboons are one of the two major primate experiments in ground-dwelling. They took to eating vegetation such as tubers, which they dug up with their forefeet and tusks. Occasionally, a troop may have killed a young or wounded animal like a rabbit or an antelope.

The other primate experiment in ground-dwelling was destined to change the tree of animal life forever. Like the baboons, this group moved into the grasslands. But unlike the baboons, this new group did not content itself with digging tubers or occasional hunting. Like the baboons, this group was composed of families of gregarious, intelligent primates. But unlike the baboons, they were not monkeys but a form more intelligent, somewhat like the great apes of today—but not apes, either, although related to them.

These were the first hominids, "manlike ones." Unlike baboons,

family of early hominids foraging (right). *Baboons and hominids are the only significant primate experiments in ground dwelling. Both have done very well so far.*

who have sharp tusks with which they protect themselves against the many predators on the grassland, the first hominids had simpler, generalized teeth without any slashing canines. To protect their families, they seem to have thrown things—much as do chimpanzees and other apes today.

Chimps, however, live in forests, as did their primate ancestors; when hard-pressed by predators, they take to the trees. On the grasslands, where there are relatively few trees, the hominids could not do this. Selective pressures from the surrounding lions, leopards, hyenas, and hunting dogs tended to favor genetic changes in hominids that permitted more accurate throwing of stones in defense, plus ever more efficient communication. As their forefeet became specialized for throwing and less suited to walking, their hindfeet became more and more adapted, through natural selection, to faster running. For the second time in the history of the tree of animal life, a two-legged runner came into being. This was *Australopithecus* ("southern ape") of Africa.

These two-legged primates, with their hooting communication and omnivorous diets, were utterly new beings in the grassland world. No animals had ever evolved defenses against accurate stone throwers, because no such stone throwers had ever lived before. Soon the hominid diet came to involve a greater and greater ability to fell small animals for food, a useful skill in a land where most of the biomass was composed of grasses that were inedible to primates. Many small animals could digest grasses, however, and the new hominids hunted them and prospered. As hominid populations increased, the smaller animals evolved to evade the agile stone throwers.

In response, the hominids coevolved. Their brains changed, perhaps because the nature of their prey required that they produce hunting strategies. They may have worked in efficient teams, some driving prey before them to others waiting hidden in the grass. Like wolves, they still retained their ability to scavenge, to eat tubers and

A group of early men, Homo erectus *("standing man"), hunting with fire, surrounds an antelope. Delight in hunting and other active sports of snatch-and-grab, especially*

fruits, to be omnivorous. And like wolves, too, they ultimately became hunters of animals larger than themselves, for with the evolution of larger brains came the capacity to share complicated strategic planning.

Now, in those days Africa (and the rest of the world) was full of huge mammals, herbivores whose size and strength protected all but the youngest and weakest from predators like wild cats and dogs. The new hominid predators, however, could circumvent this defense of size and power. They constructed tools, and compared results, and improved the tools. Tool use itself favored those hominids who possessed keen intelligence and coordination; and, brains and communication already being part of their primate heritage, these traits were enhanced.

those involving missle throwing, probably rigorously selected for among early men, survives today.

Thus was born language, and with language, gradually, was born humanity as we know it. Successive genetic changes toward greater size and adaptability, especially in the face of growing competition among the hominids themselves, forced in these hominids greater and greater dependence on their wits and speech. When groups came into competition for resources, those with more intelligence in tool use and language tended to drive less intelligent forms away or destroy them. Over the past three million years, several different hominid species have come into being, each one progressively bigger-brained, and each one usurping the econiches of the previous one.

During this time, the great mammals that the hominids hunted either became increasingly scarce or else managed to elude hominid ploys. For instance, four million years ago there were many species of elephants around the world. Today only two remain, the Indian

and the African. Both these elephant species are highly intelligent, social, and protective of their own family units, traits that must have afforded them protection from primitive human hunters. The rest of the elephants are extinct, as are most of the big mammals. By about forty thousand years ago, the human species as we know it (Homo sapiens—"man the wise") had appeared, and it drove all other hominids into extinction. Our own kind, then, inherited the earth.

Our own kind, too, inherited fire and tool-making from their ancestors. With these skills human beings were able to leave the warm climates in which they had evolved. Those most suited to the tropics, most firmly entrenched, drove marginal populations ever outward through competition. Human beings experienced a great cultural adaptive radiation as they migrated to new places and sought new foods. By about ten thousand years ago, all of the earth's continents except Antarctica were inhabited by human beings, and almost all of the large mammals were extinct. Those that remained had survived only because their speed or social organization excelled the primitive weapons of the Stone Age human beings that hunted them.

A meeting of minds at the edges of the European glaciers forty thousand years ago. A family of glacier people (Neandertalers), highly adapted to the terrible cold of the Ice Age tundras, comes upon one of the men of the warmer southlands, of a type called Cro-Magnon today. The Neandertalers, specialized to living and hunting among the ice masses, were of a heat-conservative build, with massive bones and short stature. Formerly considered primitive in structure, they were actually people of essentially modern type, with brains larger, on the average, than our own. Cro-Magnons were indistinguishable from modern Europeans except that their brains, too, were larger.

Selective pressure came yet again to bear on these people, who were by now well adapted to hunting large animals. They found those animals ever more scarce and difficult to catch. They ate all the most easily available fruits and other rich foods. Ultimately their consumption was so total, and the landscape so denuded, that huge deserts began to form. Starvation set in, and in their terrible hunger the people ate all manner of strange foods. When all else failed, they were forced to glean seeds from the very grasses, to grind those seeds with water, and to eat the slimy pastes. How hungry can you get?

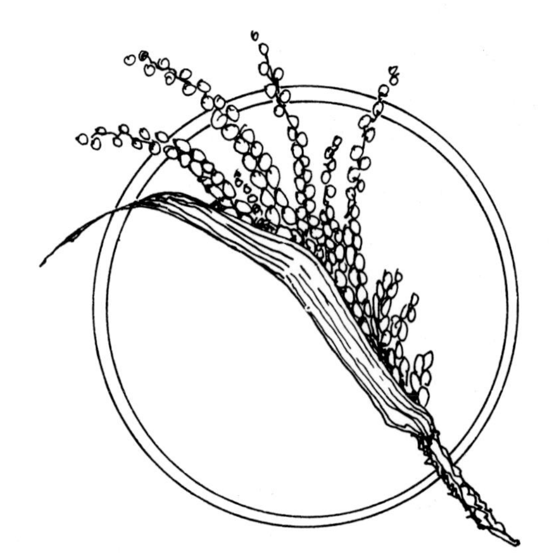

Millet, domesticated in the Near East about twelve thousand years ago.

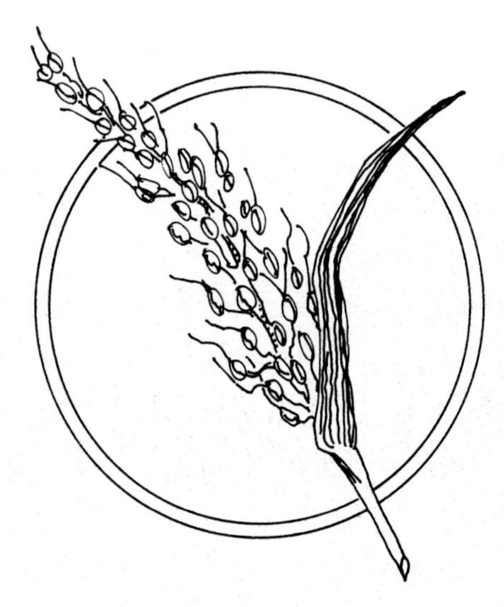

Rice, domesticated about twelve thousand years ago in the Far East.

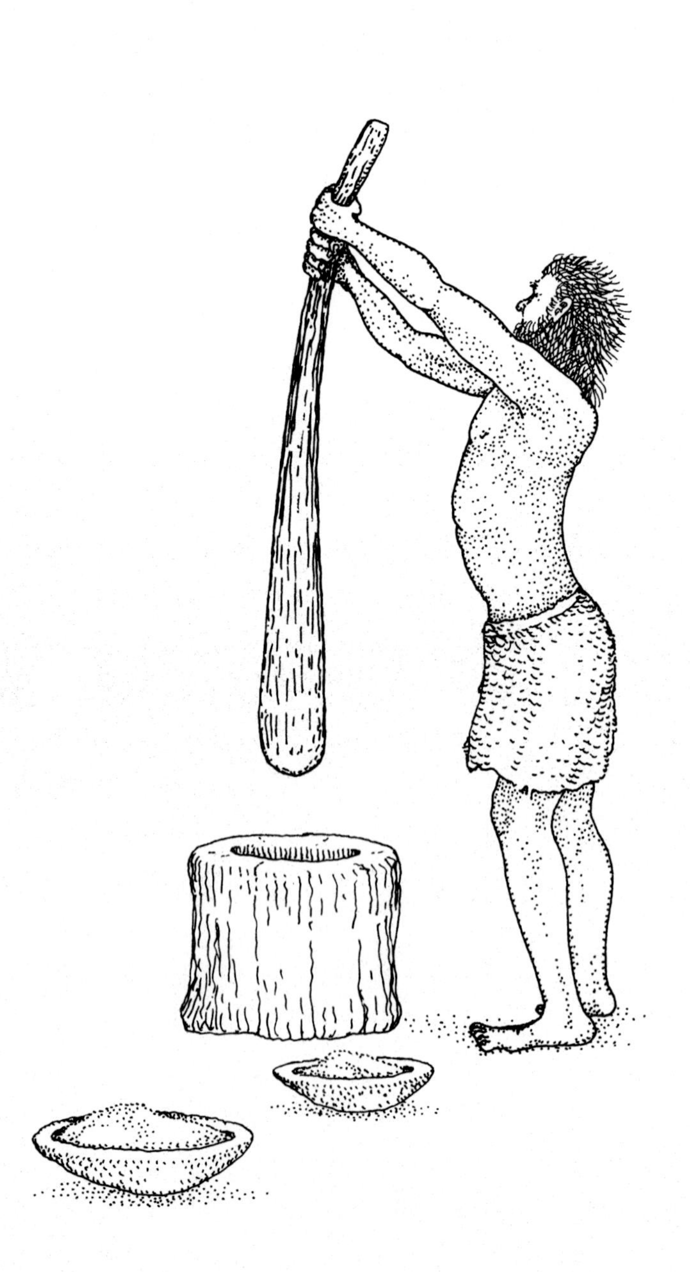

But, collecting seeds in this manner, the starving people underwent a sort of cultural mutation when they learned that such seeds were not only nourishing in themselves, but would also produce more plants, and thus more seeds, if buried in the ground. Agriculture was born, and human beings became grass-seed eaters, herbivores.

Because a carnivorous animal must capture the sun's energy from animal flesh, which is not as plentiful as the plants on which those animals feed, there tend to be fewer carnivores than herbivores in any ecosystem. This was as true for early human beings as it is now true for wolves and eagles.

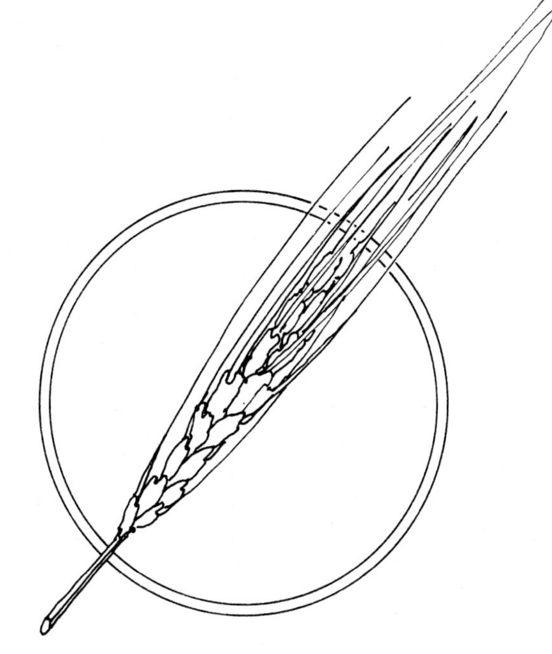

Wheat, domesticated in the Near East about twelve thousand years ago.

Teosinte, the ancestor of maize, domesticated in Central America about ten thousand years ago.

As human hunters, using fire and other strategems, killed off most of the world's big game, human beings became ever more desperate for food. The rich seeds of wild grasses served them in their time of greatest need. These grasses ultimately became domesticated—the birth of agriculture.

The rise of settled communities based on agriculture offered plenty of hunting for nomads, so agricultural peoples began building walled stockades within which they could rally during attacks.

Herbivores, on the other hand, capture the energy directly from plants. There are many herbivores to each carnivore in an ecosystem. When human beings learned to live by planting grasses such as wheat, rye, rice, and maize, separating the seeds from the chaff, and grinding the seeds into flour, they experienced a population boom that began about ten thousand years ago and continues to this day. From groups of isolated hunting families grew tribes of grass-seed eaters, and then towns of grass-seed eaters, with room besides for specialized occupations. Soldiers came into being, and generals, to protect the lands and crops and other resources on which towns de-

As agriculture improved, populations grew.

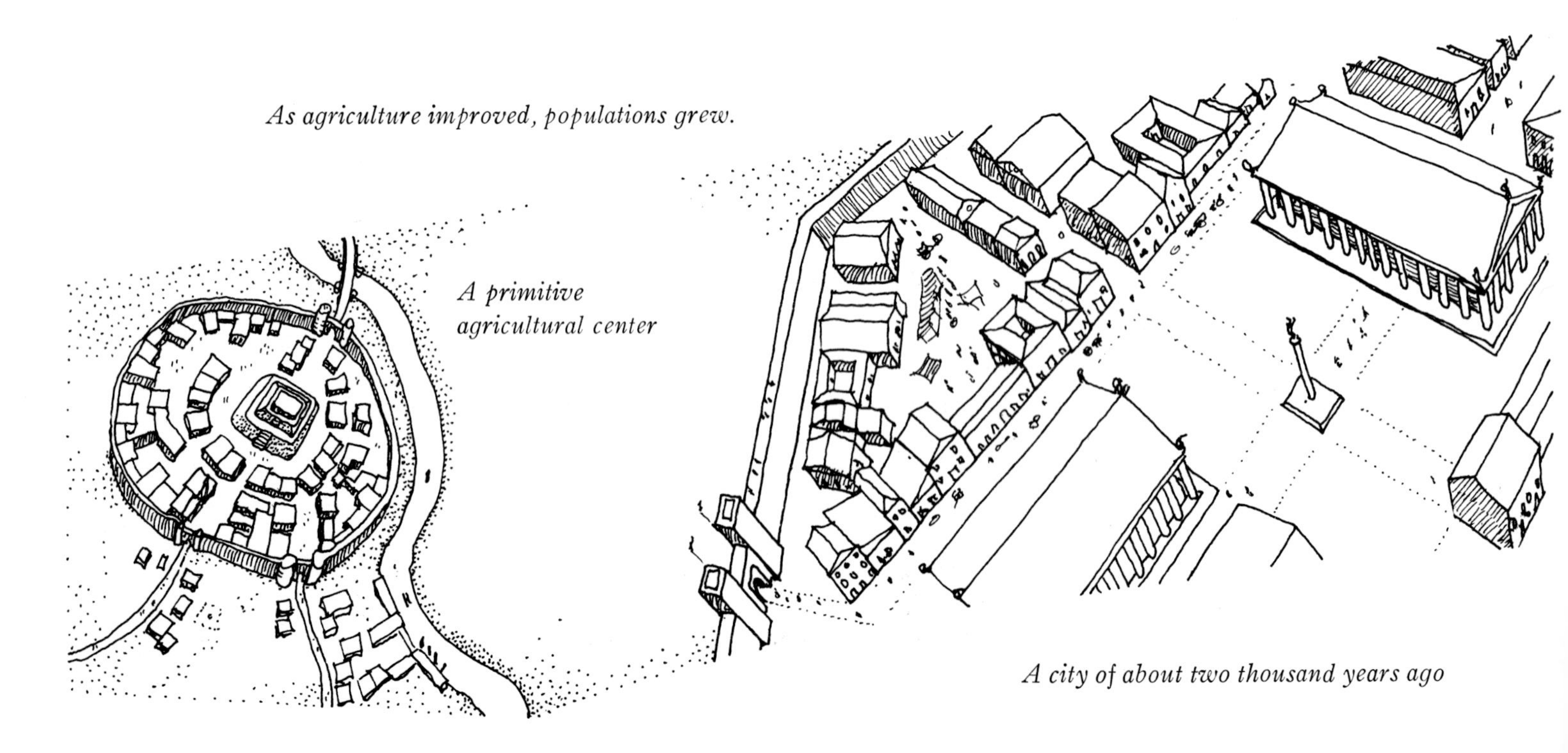

A primitive agricultural center

A city of about two thousand years ago

pend. And these soldiers could be marshaled to steal resources from other towns. As the power of towns increased, they became cities of hundreds of thousands, then nations of millions.

Now the tree of animal life is being pruned back yet again, much as it was at the end of the Mesozoic. This time, however, it is being pruned by the success of human beings, who are in the process of using up almost all of the world's living matter. Some five hundred thousand species of plants and animals may become extinct during the last fifty years of the twentieth century alone, ending the Cenozoic era forever. What will be left, we cannot say.

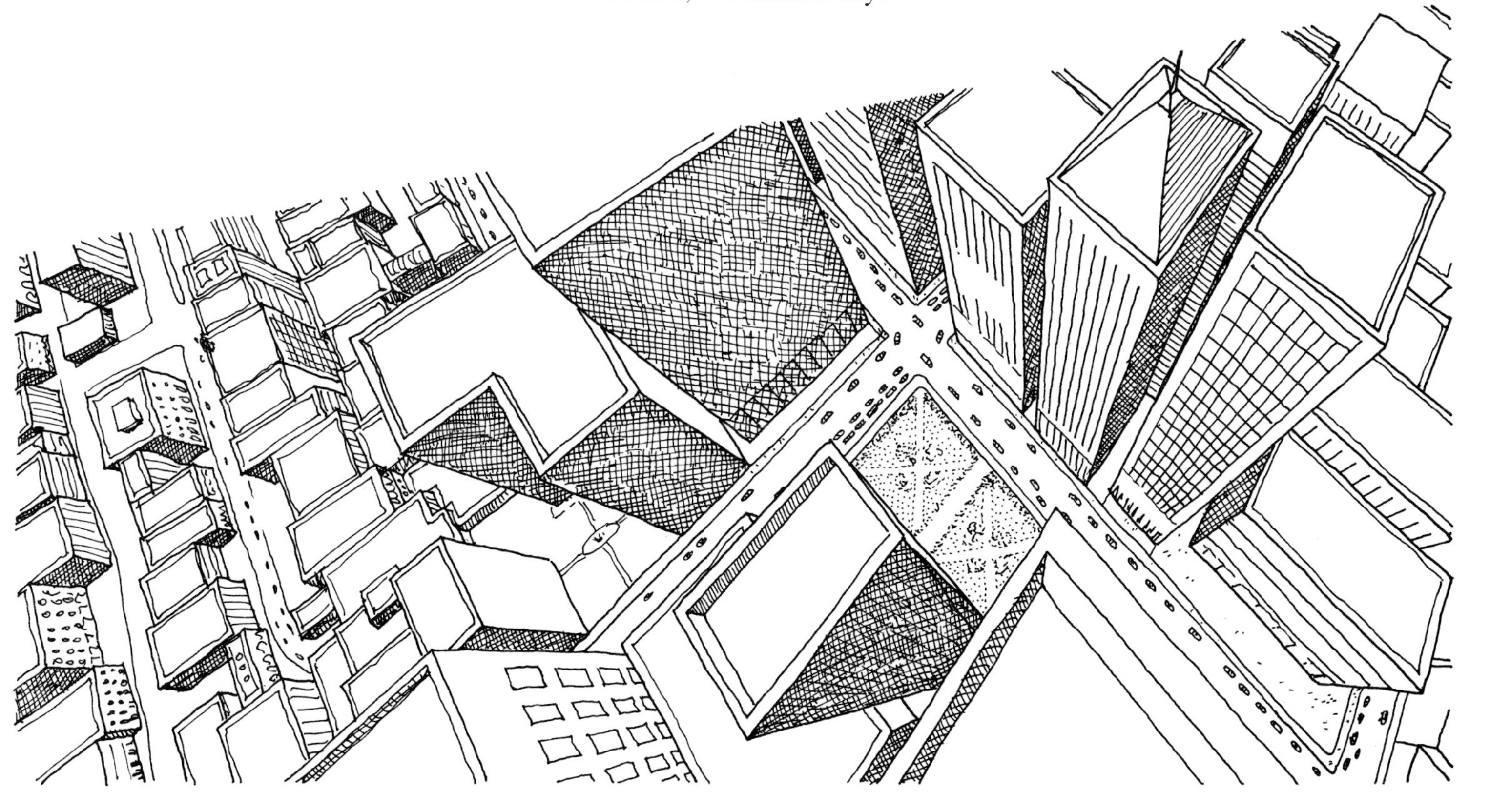

A modern megalopolis, unstable, petroleum-dependent, endangered by bankruptcy and inefficiency—but also a center of learning, of art, of science.

We do know, however, that there is less and less room on earth for human beings, as these most adaptable of animals continue to lay waste to their world. Most people alive right now will never have quite enough food for healthy living; millions are dying and will die for want of *any* food. The pressure of selection is building on us now that our great cultural adaptive radiation has taken us to the ends of the earth. In the cycle that we have seen throughout the growth of the tree of animal life, change always follows selective pressure. We now lie at the threshold of a great change, the mightiest in at least the past 65 million years.

The animal response of fear has all too often governed human affairs, producing large classes of specialized killers throughout the ages. (from left) *A Sumerian archer; an Egyptian lancer; a Roman infantryman.*

(from left) *A ninth-century Saxon swordsman; a sixteenth-century arquebusier (scatter-gunner); a twentieth-century soldier whose garb reflects the terrible spectre of chemical-biological-radiological warfare looming in our immediate future.*

Like all animals, human beings can be motivated by fear. Because they will make the great change themselves, they are confronted with the option to make this change through terror, widepsread death, and the devastation of the earth. Even now we see the accumulation of terrible weapons, thermonuclear bombs, new forms of disease and poison, rockets with artificial brains, all designed to destroy millions of human beings.

Even if we are able somehow to keep from going to war over our diminishing resources, we will have to regulate their use, and our own numbers, in order to survive on earth. The form this regulation will take is already apparent in the many computers and other information-retaining systems designed to keep track of the activities, money, and even thought of private people. Indeed, within a few decades there may be no such thing as privacy. Whether through war or regulation, the future of our overpopulated earth looks grim indeed.

The human option of hope, epitomized in the person of Charles Robert Darwin. The giant labors of this gentle philosopher and orchid breeder laid the cornerstone for our understanding of the tree of animal life, offering a supreme example of the finest capabilities of the human species.

Unlike other animals, however, human beings can also be motivated by hope. Looking into the future, perhaps they will opt to make the coming change with hope. As the only life on earth ever to have had the capability of leaving earth herself, they might even choose to initiate the next, and greatest, adaptive radiation, taking the tree of animal life into the lifeless wastes of space and saving the earth for the cherishing that she deserves from future generations. The choice is ours.

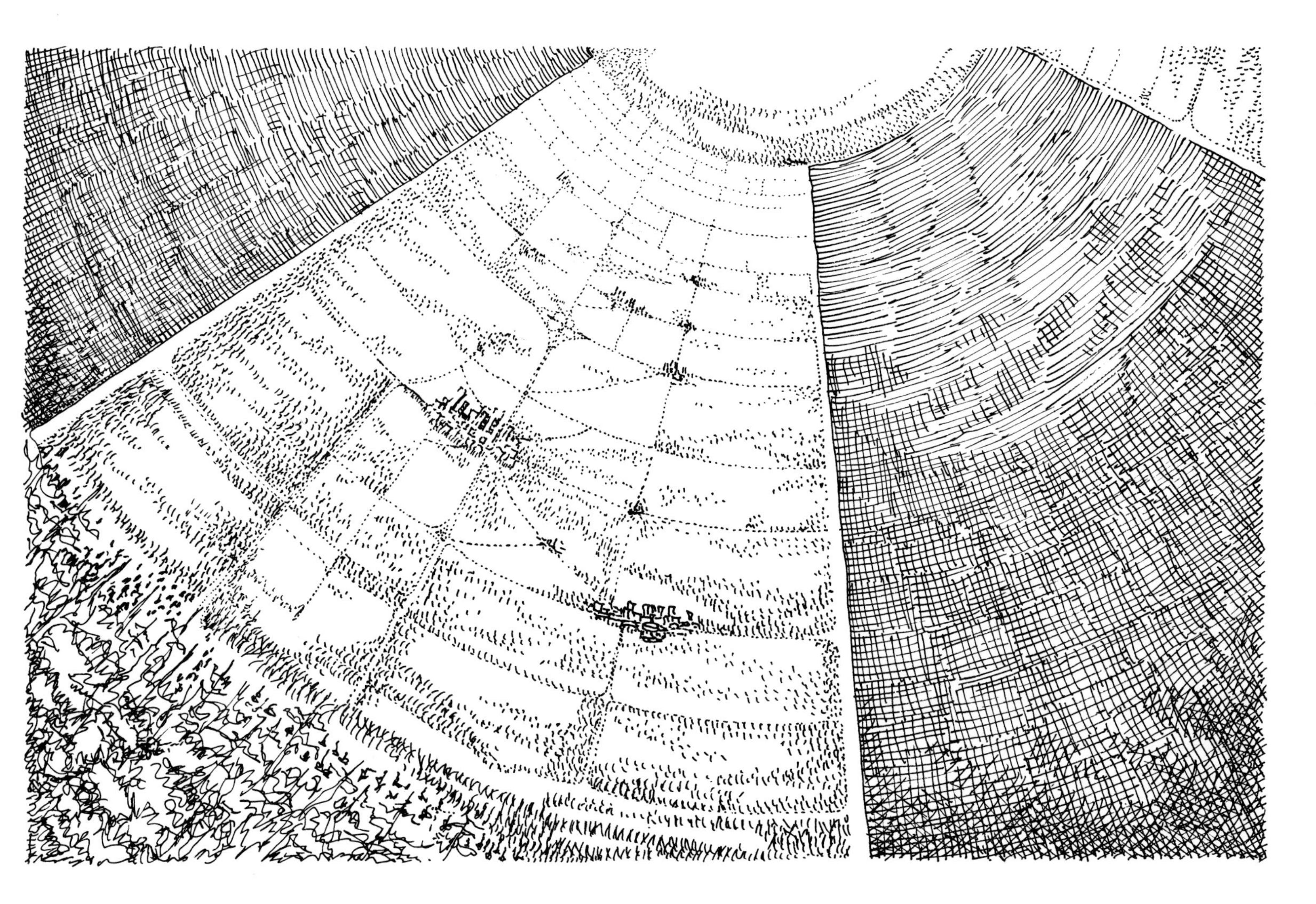

A look into a vast cylindrical colony, an entire ecosystem many kilometers long, rotating in space. Within such a structure millions of people might live, along with many other life forms, extending Earth's progeny to undreamed-of limits. Thinkers like physicist Gerard K. O'Neill have shown that these colonies could be constructed within the scope of our current technology. If we can but follow the examples of people like Darwin and O'Neill, the tree of animal life can only grow richer, more diversified, and more beautiful in the millions of years to come.

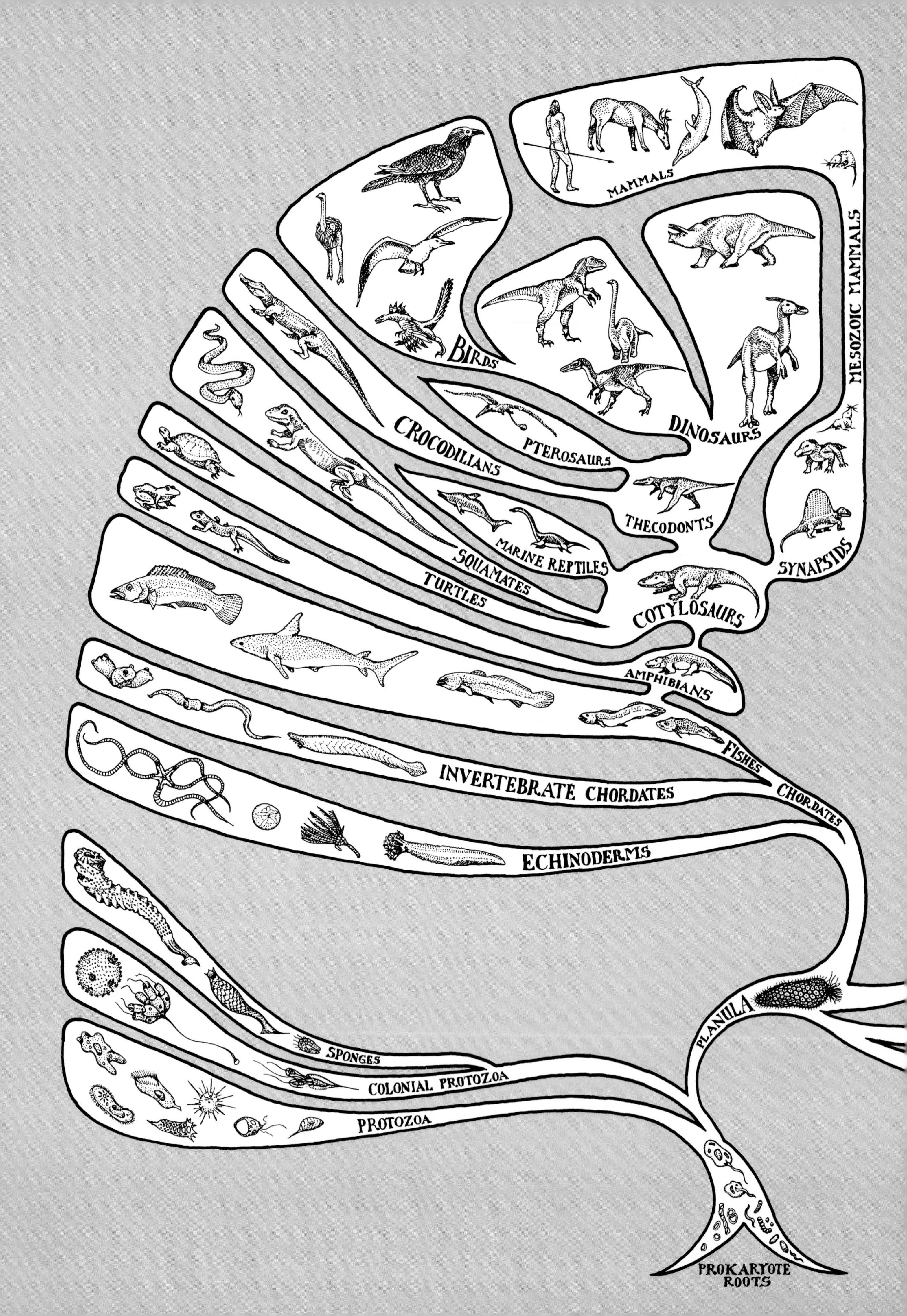
MAMMALS
MESOZOIC MAMMALS
BIRDS
DINOSAURS
CROCODILIANS
PTEROSAURS
THECODONTS
SYNAPSIDS
MARINE REPTILES
SQUAMATES
TURTLES
COTYLOSAURS
AMPHIBIANS
FISHES
CHORDATES
INVERTEBRATE CHORDATES
ECHINODERMS
PLANULA
SPONGES
COLONIAL PROTOZOA
PROTOZOA
PROKARYOTE ROOTS

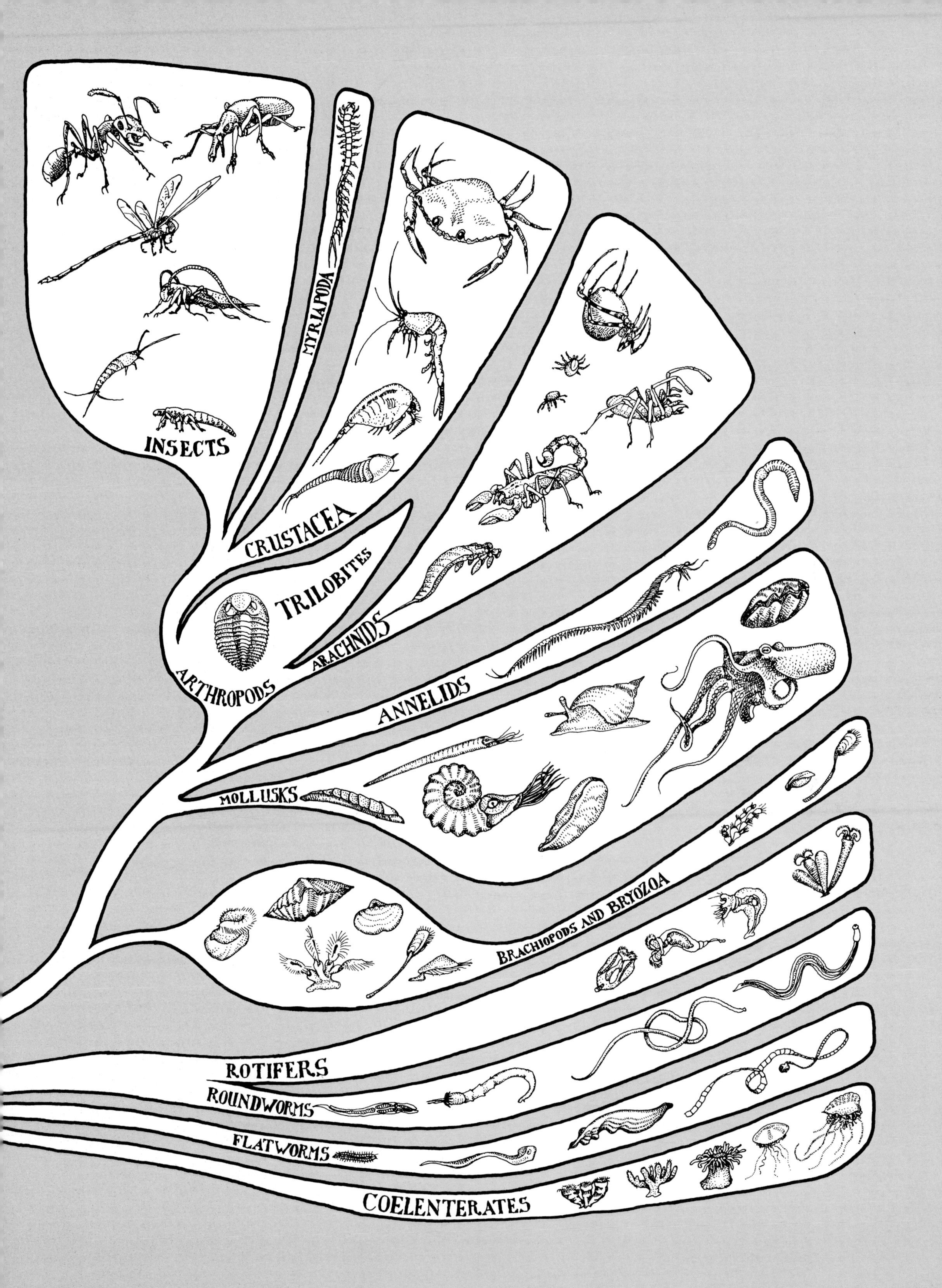
INSECTS
MYRIAPODA
CRUSTACEA
TRILOBITES
ARACHNIDS
ARTHROPODS
ANNELIDS
MOLLUSKS
BRACHIOPODS AND BRYOZOA
ROTIFERS
ROUNDWORMS
FLATWORMS
COELENTERATES

GLOSSARY

Adaptive radiation The invasion by a newly successful organism of a number of econiches unavailable to its ancestors. This eventually produces the division of a single species into many species adapted to specialized life-styles.

Aerobic Using oxygen in metabolism; said mainly of those bacteria that are able to use oxygen dissolved in water.

Algae Primitive plants. Blue-green algae are prokaryotic algae. (singular: alga)

Ammonoid A member of an extinct group of mollusks related to squid and octopuses, having a spiral shell from which its head and tentacles protruded.

Amniote Any vertebrate with eggs that contain amniotic fluid in which the embryos develop before birth. Amniotes are land dwellers, or descended from them; reptiles, birds, and mammals are familiar examples.

Amphibian A vertebrate that can breathe air, but whose young must develop in water because amphibian eggs lack amniotic fluid. Amphibia are tetrapods; they have, or are descended from amphibia that had, four legs.

Anaerobic Using no oxygen in free solution; said of bacteria that live in places containing no free oxygen.

Annelid Any member of the phylum Annelida of segmented worms (*annulus* is a "ring" of the sort encircling most of these worms). From annelids or animals very like them both mollusks and arthropods descended.

Archosaur Any member of an order of advanced amniotes that includes thecodonts, pterosaurs, dinosaurs, crocodilians, and perhaps birds, whose classification is based on certain technically linked features of skeleton and evolution.

Arthropod Any member of the largest phylum of animals, the "jointed leggers." This advanced group contains the insects, arachnids, crustaceans, and certain extinct forms such as trilobites and sea-scorpions. Arthropods have an external skeleton with jointed appendages.

Autotroph An organism that can manufacture its own food from inorganic materials, water, and some energy source such as sunlight.

Bacteria Prokaryotic cells that usually do not carry out photosynthesis. (singular: bacterium)

Biomass The total weight of living matter in a species, ecosystem, or other designated living community.

"Bird-hipped" dinosaur A dinosaur whose hipbones fancifully resemble a bird's in their positioning. These dinosaurs comprise the order Ornithiscia, which means "bird-hipped"; all were herbivorous.

Bony fish A fish whose skeleton is composed of bone rather than cartilage as in sharks.

Brachiopod Any member of the phylum Brachiopoda of filter-feeding clamlike animals whose shell halves lie above and below the body rather than side by side as in clams.

Bryozoan "Moss animal," any member of the phylum Bryozoa of small filter-feeding animals with mouths surrounded by a ridge of ciliated (hairy) tentacles.

Burgess Shales Fossil-bearing rocks deposited about 550 kilometers east of Vancouver, British Columbia. The well-preserved fossils in this formation give us a good look at life of around 530 million years ago.

Carnivore Any animal that kills and eats other animals.

Cenozoic The "Era of the New Animals," during which mammals became the dominant vertebrates of the earth's landmasses. It began about 65 million years ago and has lasted to the present.

Chordate Any member of the phylum Chordata of animals which at some time during their lives possess notochords, stiffening rods extending the lengths of their bodies. Vertebrates, including ourselves, are members of this phylum.

Chromosome A "colored body," any of the rod-shaped bodies in the nuclei of eukaryotic cells that contain genetic information.

Class In taxonomy, the grouping of organisms immediately smaller than a phylum in size.

Coelenterate "Hollow-innard," any member of the phylum Coelenterata of simple metazoans whose bodies consist of a hollow gut surrounded by a muscular body. Jellyfishes, corals, sea anenomes, and hydras are common coelenterates.

Convergence An evolutionary process in which organisms occupying similar econiches tend to acquire similar characteristics.

Cotylosaur "Stem lizard," any of the first amniotes.

Dinosaur Any member of two orders of archosaurs including the dominant land animals of the Mesozoic era. Dinosaurs were highly advanced and active creatures descended from thecodonts. Some taxonomists have proposed including the dinosaurs and birds within their own class Dinosauria to differentiate them from the reptiles with which they are traditionally grouped.

DNA *D*eoxyribo*n*ucleic *a*cid, the large spiral molecule along which the code, or instructions, for the growth of a living organism is arranged.

Ecology The study of the distribution by the living system of matter, energy, space, and time.

Econiche In the study of ecology, an organism's unique "job," or function.

Ectotherm An animal that must depend on external temperature conditions to maintain its internal temperature; cold-blooded.

Edentate Any member of an order of mammals that evolved in South America when it was an isolated continent, including sloths and armadillos; the word means "no teeth," although most have teeth (plenty of them).

Ediacaran Any rock or fossil from the Ediacaran hills of Australia; the Ediacaran fossils provide a glimpse of animal life between 600 and 700 million years ago.

Egg The female sex cell, containing the larger reserve of food for the young.

Endotherm An animal whose body temperature is maintained by internal metabolic processes rather than being dependent on outside temperatures; warm-blooded.

Eukaryote An advanced cell with a nucleus.

Eurypterid A sea-scorpion, a predatory aquatic arthropod of Paleozoic times. Sea-scorpions became extinct as vertebrates arose to dominate large animal econiches in the water.

Evolution The process by which the forms and functions of living beings change from one generation to the next in response to changes occurring in their environments and in their genetic material.

Extinction The total dying out of a species or larger group of organisms.

Fossil From the Latin for "dug up"; any trace of a living form that existed in the past.

Gene Information stored on the chromosomes in the nuclei of an organism's cell or cells that is hereditary—it can be passed on to the young. Genes determine an animal or plant's characteristics in life.

Herbivore Any animal that eats plants for a living.

Heterotroph An organism that must consume other organisms or dissolved energy-rich molecules for food.

Hominid Any member of the family Hominidae, bipedal, tool-using primates including human beings, the only living representatives.

HOMOLOGOUS Physical structures in different animals that are similar on account of having evolved from a single structure in a common ancestor (as the wing of a bird and the arm of a person).

ICHTHYOSAUR "Fish-lizard," an amniote highly adapted to the marine world and resembling a dolphin in form. Ichthyosaurs became extinct before the end of the Mesozoic era.

INSECT Any member of the class Insecta, six-legged arthropods with bodies divided into three sections: head, thorax, and abdomen. Insects are by far the largest class of animals in terms of number of species. They evolved on land and remain for the most part land animals.

INSECTIVORES Any member of an order of primitive placental mammals that eat insects, including shrews, moles, and their relatives.

INVERTEBRATE Any animal that is not a member of the subphylum Vertebrata.

"LIZARD-HIPPED" DINOSAUR A dinosaur whose hipbones fancifully resemble those of a lizard. These dinosaurs comprise the order Saurischia, "lizard-hipped," and include all carnivorous dinosaurs and certain long-necked herbivores, most of which walked on all fours.

MAMMAL Any member of the class Mammalia, amniotes that feed their young with milk from mammary glands. Mammals are endotherms; they heat themselves internally, and thus most are insulated by fur, fat, or clothing.

MARINE REPTILE Any amniote adapted to existence in seas. During the Mesozoic era marine reptiles occupied the econiches now occupied by whales, seals, and other aquatic mammals. A few large sea-turtles and certain snakes remain, too.

MARSUPIAL Any mammal of the subclass Metatheria whose young are born at an early stage of development and must be carried about by their mother in a pouch until they can move about on their own.

MESOZOIC The "Era of the Middle Animals," the time of the ascendancy of the dinosaurs, lasting from 230 to 65 million years ago. The Mesozoic ended suddenly and inexplicably with the extinction of most of the world's advanced animals.

METAZOAN Any multi-celled animal.

MOLLUSKS Any member of the phylum Mollusca of soft-bodied animals including squids and octopuses, clams, oysters, chitons, snails, slugs, and the like. The mollusks are the second largest phylum, after arthropods.

MONOTREME A primitive mammal that lays eggs. Duckbilled platypuses and echidnas of the Australian region are the only living monotremes.

MUTATION Any inheritable change in the genetic material that alters the structure and function of an organism.

MYRIAPOD Any member of an order Myriapoda, "many-legged" land-dwelling arthropods that include centipedes, millipedes, and symphylans.

NATURAL SELECTION The process in which environmental conditions act to kill off organisms that are in some way unfit to reproduce as successfully as their fellows. Over many generations, a line of organisms becomes more and more precisely suited to a particular econiche as the better adapted members of the population tend to survive and reproduce.

PALEOCENE "Old Dawn of the Recent," the earliest epoch of the Cenozoic, lasting from 65 to 55 million years ago.

PALEONTOLOGY The study of extinct forms of life.

PANTODONT Any of the mammals of the Mesozoic from which marsupial and placental mammals evolved. Pantodonts were small shrewlike creatures with long snouts; they may have laid eggs.

PANGAEA "All land," a name coined by geologists for the ancient supercontinent of which the modern continents are fragments that have drifted apart through hundreds of millions of years.

PHYLUM The greatest taxonomic grouping under kingdom.

Placental Having a placenta; said of mammals whose young, nourished for a long time within the body of the mother, are born at an advanced stage of development because of the mother's placenta, an organ that allows exchange of food and waste between the blood of the mother and offspring.

Placoderm A class of primitive "plate-skinned" fishes whose bodies were armored with plates of bone and whose skeletons underneath were largely cartilaginous. This is the only class of vertebrates yet to have become completely extinct.

Planula An early metazoan form consisting of an aggregation of cells with a "head" of sensory cells and a "tail" of swimming cells; the planula lives on as the larval form of coelenterates and lies close in form to the root of the metazoan adaptive radiation. (plural: planulae)

Predator An animal that hunts other animals for food.

Prey An animal that is hunted for food by a predator.

Primate Any member of the order Primates, generalized omnivorous mammals whose ancestors invaded treetops during late Mesozoic times and whose living representatives include human beings, apes, monkeys, lemurs, tarsiers, and (some say) tree-shrews.

Prokaryote A primitive cell without a nucleus.

Protein One of the most fundamental substances, or building blocks, of living organisms, composed of amino acids.

Protozoan A single-celled animal.

Pterosaur A "winged lizard" of the incorrectly named group of archosaurs whose wings, made of leathery skin often covered by fur, stretched from their bodies to the tips of their fourth fingers. Pterosaurs arose early during the Mesozoic but were largely replaced by birds, whose feathered wings were more durable. The last of the pterosaurs died at the end of the Mesozoic.

Replication Duplication, of an organism or anything else.

Reptile A "crawler"; any member of the class of low-slung amniote ectotherms, traditionally including turtles, snakes, lizards, and crocodilians. Many taxonomists also include such non-crawlers as dinosaurs within the class.

Squamate Any member of the order Squamata of reptiles, "scaled ones," including snakes and lizards.

Specialization The process in which an organism becomes adapted to a narrow econiche, while losing its options in the wider scheme of things.

Sperm The male sex cell, usually the one that seeks the female because it is lighter and requires little energy to get about.

Synapsid Any member of a group of advanced amniotes from which mammals descended. Synapsids were characterized by a large opening at the rear of the skull within which the jaw muscles moved. Because this opening was formerly believed to have been fused from two others, the term synapsid (fused-arched) was coined for the group.

Thecodont Any member of a group of primitive crocodile-like archosaurs from which both true crocodiles and the more advanced dinosaurs descended. The name means "socket teeth."

Therapsid Any member of the higher synapsids sporting comparatively long legs, advanced differentiated teeth (incisors, canines, molars), and other mammal-like characteristics, of which mammals are descendants.

Trilobite A primitive arthropod having a body of many similar segments. The name means "three-lobed stone." Trilobites died out before the end of the Paleozoic and were replaced by more advanced arthropods, mainly crustaceans.

Vertebrate A chordate whose nerve cord is surrounded by rings of bone or cartilage. These rings are called vertebrae, and vertebrates occupy a subphylum Vertebrata of their own. Fishes, amphibia, reptiles, birds, and mammals are familiar vertebrates.

INDEX

Illustrative material as well as text has been included in the index.